INTRODUCTION TO BIOLOGY I and II

JASON B. JENNINGS

Southwest Tennessee Community College

Kendall Hunt
publishing company

Kendall Hunt
publishing company
www.kendallhunt.com
Send all inquiries to:
4050 Westmark Drive
Dubuque, IA 52004-1840

ISBN 978-1-4652-0196-6

Printed in the United States of America
10 9 8 7

CONTENTS

PREFACE

This edition of *Introduction to Biology I and II* is intended for non-science majors and includes labs suitable for a two-semester course in introductory biology. The purpose of this manual is to generate interest in biology by those students who might otherwise be intimidated by the subject. Therefore, it is written in a style that is easily understood by today's typical undergraduate student. It includes labs regarding basic biological principles: chemical bonding, cell structure and processes, photosynthesis, cellular respiration, genetics, anatomy, surveys of the different kingdoms, natural selection, and other topics. The labs can be completed in a single lab period using equipment that can be found in most college biology laboratories. Individual labs feature varied questions throughout the activity designed to test the student's understanding of the concept. Lastly, there are post-lab question sheets that may be removed and turned in for grading.

EXERCISE 1

THE SCIENTIFIC METHOD

INTRODUCTION

One of the differences between scientists and non-scientists is the way the two groups solve problems and answer questions. Scientists typically use the scientific method (Figure 1.1) to answer questions. The scientific method consists of a series of steps that always begins with some observation. You observe something in nature, and it captures your interest. For instance, you may notice a group of birds such as crows attacking a hawk. This type of behavior, called "mobbing," is quite common. When you make an observation you are likely to ask questions concerning what you saw. For the mobbing observation you might wonder why the crows are harassing the hawk. Once you have asked a question, the next step in the scientific process begins.

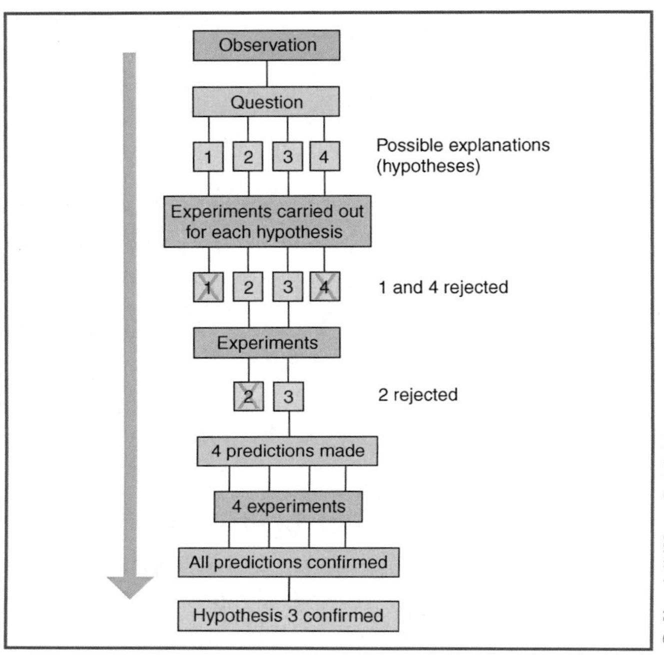

FIGURE 1.1

Several possible explanations may exist for any given question. These are called **hypotheses**. Hypotheses must be testable and make predictions. In the previous example several hypotheses may explain this behavior by the crows. Could the crows engage in this mobbing behavior because they view the hawk as possible food? Could the crows engage in this behavior because they view the hawk as a possible competitor for the same food source and are trying to drive it away? Are the crows mobbing this hawk because they believe the hawk will attack their nest if they do not drive it away? These are all possible explanations, but how do you know which one (if any) is correct? The next step in the scientific process is experimentation, and it will address this problem. **Experiments** are designed in such a way that once you complete one you have supported one hypothesis over all the others. It is important when you design an experiment to set it up as a **controlled experiment**.

CONTROLLED EXPERIMENTS

Controlled experiments divide your test subjects into two or more groups; one is the **control group**, and the others are the **treatment groups**. You then compare these two groups to determine if there are any differences between them once the experiment has concluded. In any experiment you will need to consider a number of variables.

Independent variable: This is the variable you change between the control group and the experimental group.

Dependent variable: This is the variable you are interested in observing.

Controlled variable: These are variables that are the same between all groups (the control and the experimental group). This is an important point. Excluding the independent and dependent variables, all other variables must be the same between groups. In this way you can safely say that any differences between the control and experimental groups with respect to the dependent variable are because of the independent variable.

Sound confusing? Let's look at a simple example. Suppose you made a drug that you thought cured the common cold. How would you go about testing it? The first thing you would have to do is to formulate your hypothesis, which might be something like this: I believe that this drug will cure those patients of the common cold who take it. Your next step is to design your experiment. You will need two groups: a control group and an experimental group. In this case your independent variable is the treatment (your drug and a placebo: sugar pill). The experimental group will receive your pill, and the control group will receive the placebo. This way you can measure how long it takes your subjects to get over the common cold (dependent variable). How do you assign people to the different groups? This addresses the controlled variables. The two groups must be as similar as possible (same general ages, health, etc.) so you can attribute any differences between the two groups to your drug. Make sure the subjects do not know which group they belong to. You treat the two groups exactly the same except one group gets the pill and one group gets the placebo.

ANALYZING YOUR DATA AND DRAWING CONCLUSIONS

This is the next step in the scientific method. In the previous example your results would be how long it took to cure the two groups of the common cold. For example: The experimental group (the one taking your pill) had no symptoms of the common cold twelve hours after taking your medication, and those symptoms did not return. The control group patients remained sick for five days before getting better. In this case you have supported your hypothesis because those people who received your drug got better faster than those that received the placebo. However, suppose those that took your drug did not get better. Instead, those that took the placebo and those that took the drug stayed sick for five days. In this situation you failed to support your hypothesis, so you can conclude your drug does not cure the common cold. In this case, did your experiment fail? NO. In science there are no failures. While you failed to support your hypothesis, you still learned something, and that is what science is all about.

PLANT GROWTH EXPERIMENT

Take a look at the data chart below and answer the questions that follow. These are data from a hypothetical plant growth experiment. Plants are in 3 groups with 10 plants each. Plants in group 1 grew in soil with no fertilizer added. Plants in group 2 grew in soil with added nitrogen. Plants in group 3 grew in soil with added potassium. Plant growth measurement is in centimeters and refers to the height of the plant.

PLANT #	GROUP 1	GROUP 2	GROUP 3
1	15 cm	23 cm	24 cm
2	12	21	27
3	8	32	28
4	6	29	31
5	10	27	30
6	12	33	29
7	5	30	28
8	14	28	30
9	11	31	29
10	12	34	31

1. Based on the data, what do you think the investigator's hypothesis was?

2. Calculate an average growth for each of the 3 groups and graph the results in the space provided.

3. Based on these results, was the investigator's hypothesis supported?

4. Why do you think the investigator used 10 plants per group instead of 1 plant per group?

5. What was the purpose of group 1?

6. Can you think of any other parameters that you can measure other than the height of the plant?

7. If you were to do this experiment over would you do it differently or the same? Why?

MOBBING EXPERIMENT

Remember the observation made at the beginning of the lab concerning mobbing? How would you design an experiment to address this problem? Work as a lab group to design an experiment to do just that. Remember to include your hypothesis, variables, etc. Also, remember that there may not be one right way to set this experiment up. Be creative and use the scientific method.

NAME ______________________________

EXERCISE 1

QUESTIONS

1. What is the purpose of a control group?

2. What is an independent variable?

3. What is a dependent variable?

4. What is a hypothesis?

5. What would you do if you failed to support your hypothesis?

6. What are the requirements for a good hypothesis?

EXERCISE 2

METRIC MEASUREMENT

INTRODUCTION

In 1790 the National Assembly of France requested the French Academy of Science to "deduce an invariable standard for all measures and weights." The French Academy responded to the request by creating the Metric System. Today, a modernized version of the metric system, established by international agreement, provides a logical and interconnected framework for all measurements in science, industry, and commerce. This International System of Units, abbreviated SI, is built upon a foundation of eight base units. All other SI units are derived from these units. Multiples and submultiples are expressed in a decimal system.

The great advantage of the metric system is its simplicity. You will see that the metric system is easy to use because all conversions are made with multiples and submultiples of 10. In order to use the metric system, you will need to know the meaning of the following multiples, prefixes, and symbols.

The eight base units are listed in the following table.

BASE UNIT	SYMBOL	USED IN MEASURING
METER	m	Length
KILOGRAM	kg	Mass
SECOND	s	Time
AMPERE	A	Electric current
KELVIN	K	Temperature
MOLE	mol	Amount of substance
CANDELA	cd	Luminous intensity
LITER	l	Volume

In this exercise, you will become familiar with the metric system and its uses. You are already familiar with some metric units. The base units for time, electric current, amount of substance, and luminous intensity are the same as you have used in the English System.

It is important that you become familiar with the metric system. The United States is in a transitional period in which the metric system will eventually replace the English system and become the only system used in science, industry and commerce.

MULTIPLES AND SUBMULTIPLES		PREFIXES	SYMBOLS
1 000 000 000 000 000 00	$= 10^{18}$	EXA	E
1 000 000 000 000 000	$= 10^{15}$	PETA	P
1 000 000 000 000	$= 10^{12}$	TERA	T
1 000 000 000	$= 10^{9}$	GIGA	G
1 000 000	$= 10^{6}$	MEGA	M
1 000	$= 10^{3}$	KILO	k
100	$= 10^{2}$	HECTO	h
10	$= 10^{1}$	DEKA	da
1	$= 10^{0}$		
0.1	$= 10^{-1}$	DECI	d
0.01	$= 10^{-2}$	CENTI	c
0.001	$= 10^{-3}$	MILLI	m
0.000 001	$= 10^{-6}$	MICRO	μ
0.000 000 001	$= 10^{-9}$	NANO	n
0.000 000 000 001	$= 10^{-12}$	PICO	p
0.000 000 000 000 001	$= 10^{-15}$	FEMTO	f
0.000 000 000 000 000 001	$= 10^{-18}$	ATTO	a

Thus a kilometer is 1000 meters, a millimeter is 0.001 meters and a centimeter is 0.01 meters. The above prefixes are used with all metric units used in measuring length, volume, and mass [the meter (m), liter (l), and gram (g), respectively].

Conversions within the Metric System

To convert a measurement to a smaller unit (e.g., grams to milligrams) one simply multiplies by units of ten. In this case, multiply by 1000 and move the decimal three places to the right. Thus:

$$2.0\text{ g} = 2000\text{ mg}$$

$$5.01 = 5000\text{ ml}$$

When converting a measurement to a larger unit (e.g., centimeters to meters) divide by units of 10 (divide by 1000 and move the decimal two places to the left). Thus:

$$3.0\text{ cm} = 0.03\text{ meters}$$

Note that:

1. 1 m = 10 dm = 100 cm = 1000 mm = 1,000,000 = mm

thus:

1.2192 m = 12.192 dm = 121.92 cm = 1219.2 mm = 1,219,200 = μm

and

2. 1,000 mm = 100 cm = 10 dm = 1 m

thus:

8.2 mm = 0.82 cm = 0.082 dm = 0.0082 m

Converting measurements from the English System to the Metric System will require the use of the appropriate conversion factor. Many of the conversion factors you will need are given below.

Conversion Factors for Length

1 inch = 2.54 cm
1 foot = 30.48 cm
1 mile = 1.609 km
1 centimeter = 0.3937 in
1 meter = 3.281 ft
1 kilometer = 0.6214 mi
Millimeter: 0.001 meter = approximately the diameter of paper clip wire
Centimeter: 0.01 meter = approximately the width of a paper clip.

Conversion Factors for Volume

1 liter = 1.057 liquid quarts
1 liquid quart = 0.9463 liter
1 gallon = 3.785 liters
1 cubic foot = 28.32 liters
1 cubic inch = 16.39 cc or ml
1 fluid ounce = 29.57 cc or ml
1 tsp = 4.9 ml

Conversion Factors for Mass

1 pound (avoir.) = 453.6 g
1 ounce (avoir.) = 28.35 g
1 kilogram = 2.205 lb (avoir.)
1 gram = 15.43 grains
1 gram = 0.320 ounces

Conversion Factors for Area

Hectare = 10,000 m^2 (2.74 acres)
1 acre = 0.407 hectares

Conversion Factors for Energy

Calorie = amount of heat required to raise the temperature of 1 gram of water 1°C.
Kilocalorie = 1,000 calories
BTU = amount of heat required to raise the temperature of 1 pound of water 1°F.

OBJECTIVES

After completing this exercise the student should be able to:

A. Use the Metric System in making measurements of length, volume, and mass.

B. Make conversions within the Metric System, from the Metric to the English System, and from the English System to the Metric System.

PROCEDURE

A. Linear Measurements

1. Measure the length of your pencil in centimeters. Convert that measurement to millimeters and then to inches. The length of your pencil is ________ cm, ________ mm, or ________ inches.

2. a. Obtain a cork borer. Cut a core sample from a potato that is 5–8 millimeters in diameter and approximately 2.0 cm long. Follow your laboratory instructors instructions in using the cork borer.

   b. Square both ends of your potato core with a razor blade and measure its length in millimeters. Convert this measurement to centimeters, then inches, and finally feet. The length of your potato core is ________ mm, ________ cm, ________ inches and ________ feet.

B. Volumetric Measurements

1. Measure the volume of a small solid object by the fluid displacement method.

   a. Obtain a small 10 ml graduated cylinder and with an eyedropper add 5.0 ml of water until the fluid meniscus is exactly on the 5.0 ml line.

   b. With a forceps, gently lower your potato core into the graduated cylinder and record the new fluid level.

   c. The difference between the initial 5.0 ml volume and your final volume is the volume of fluid displaced by the potato core. The volume of your potato core is ________ ml.

2. Measure the volume of an empty soda can.
   a. Fill the can with water.
   b. Pour the water from the can into a 500 ml graduated cylinder.
   c. The volume of the can is ________ ml. Convert this volume to fluid ounces. The volume is ________ ounces. Do these values agree with the volume given on the can? ________
   d. Can you explain why your measured volume differs from the volume given on the can?

C. Measurement of Mass

1. Place a piece of paper on the balance pan and record the weight of the paper. The weight of the paper is ________ grams.
2. Carefully blot your potato core with a paper towel to remove the excess moisture.
3. Place the core on the paper and record the total weight of the core and paper to the nearest hundredth of a gram. The total weight is ________ grams.
4. Determine the weight of the potato core by subtracting the weight of paper from the combined weight of the core and paper. The weight of the potato core is ________ grams.
5. Convert the weight of the potato core to ounces then pounds. The weight of the potato core is ________ ounces or ________ pounds.

D. Temperature

A thermometer will allow you to measure temperature as degrees Fahrenheit (°F) or Centigrade (°C). Conversion between these scales is achieved by using the following simple formulae:

$$°F = (°C \times 1.8) + 32°$$

$$°C = \frac{(°F - 32°)}{1.8}$$

Using either a centigrade or fahrenheit thermometer, make the following temperature measurements and conversions.

Room (air) ________ °F = ________ °C

Tap water (cold) ________ °F = ________ °C

Tap water (hot) ________ °F = ________ °C

MATERIALS

Potatoes; cork borers (5–8mm dia); rulers with metric and English units, 10 ml graduated cylinders, distilled water, forceps, empty soda cans, balances, thermometers (centigrade and fahrenheit), 500 ml graduated cylinders, razor blades.

NAME ______________________________

EXERCISE 2

QUESTIONS

1. A ten mile bicycle ride would be _______ km; _______ m; _______ cm; _______ mm.

2. The temperature during a recent summer reached 110°F. What temperature would the TV weather guide report if he used the celsius (centigrade) scale?

3. A boy 6′ in height has a girl friend who is 5′2″. What would be their respective heights in cm?

4. Our football team's star defensive player weighs 199 pounds. He has a mass of _______ kg.

5. A man left St. Joseph, Missouri with a full tank of gasoline and headed for Indianapolis, Indiana via U.S. Highway 36. After traveling 406 miles, he stopped in Rockville, Indiana, and filled his tank with 80 liters of gasoline. (Indiana gas stations use the metric system!) How many miles did his car get per liter of gasoline?

6. How many miles per gallon did the car in the previous problem get?

7. In the previous problem, how many kilometers did the man's car get per liter of gasoline?

8. According to a road sign, the distance from Rockville to Indianapolis is 37 kilometers. How many liters of gasoline can the man in the above problem expect to use during his trip from Rockville to Indianapolis? How many gallons can he expect to use during his trip from Rockville to Indianapolis?

EXERCISE 3

THE MICROSCOPE

INTRODUCTION

Most people are aware of many of the living organisms around them everyday because they can see them. However, there is another world to explore that we cannot see with the naked eye. These organisms are **microscopic**; that is, we can only see them with the aid of a **microscope**. Microscopes have come a long way since the days of Anton van Leeuwenhoek (ca. 1600s). The types of microscopes found in laboratories today include stereoscopes (Figure 3.1), scanning electron microscopes and the compound light microscope (Figure 3.2), which we will learn to use today.

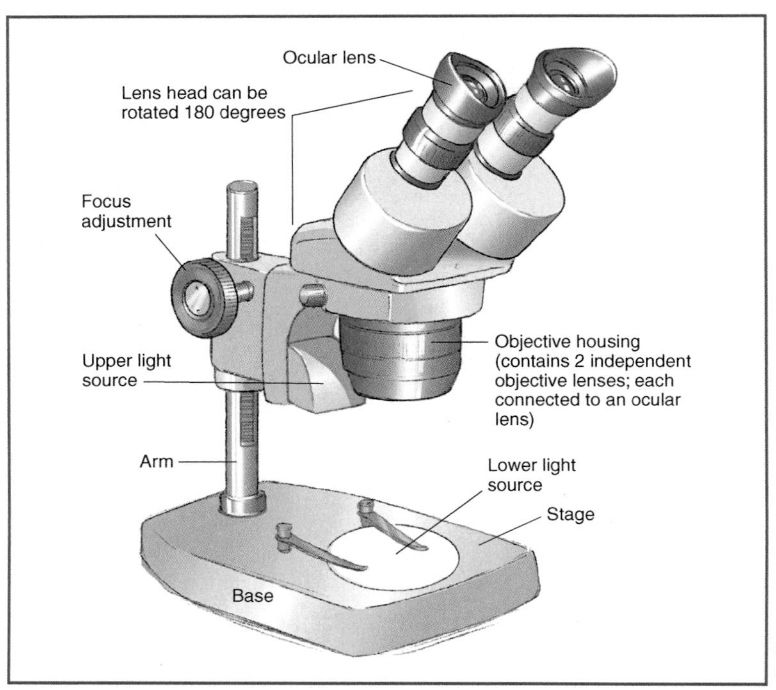

FIGURE 3.1

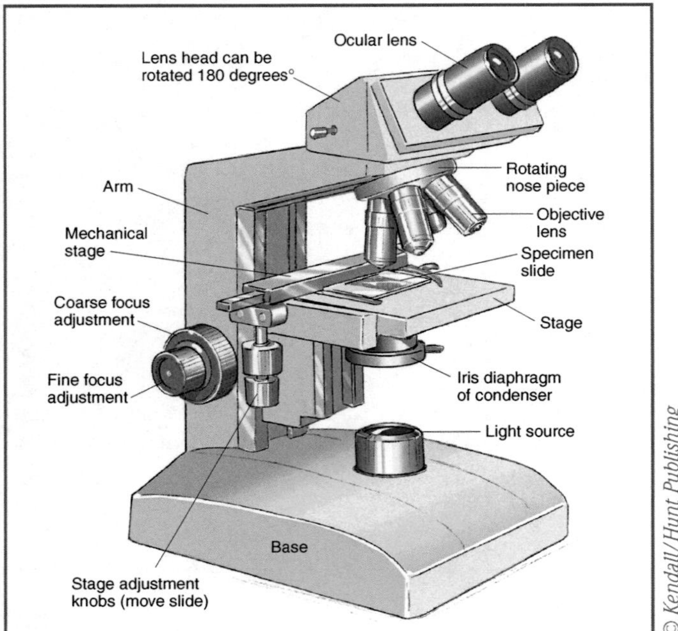

FIGURE 3.2

PARTS OF THE COMPOUND MICROSCOPE

a. **Ocular (eyepiece).** This is the part you look through. The ocular itself indicates its magnifying power. Usually the magnifying power of the ocular is 10×.

b. **Body tube.** This is the part of the microscope that attaches to the oculars.

c. **Head.** This attaches to the tube and can rotate.

d. **Arm.** This structure attaches to the head and is one of the parts of the microscope that you should use to carry it. Always carry the microscope with one hand on the arm.

e. **Rotating nosepiece.** This is the structure you attach to which the objectives are attached to. You can rotate it to change the objective.

f. **Objectives.** Objectives have magnifying power, indicated on the objective itself. The smallest objective has the lowest power while the largest objective has the highest magnifying power. Objectives are very delicate, and you should only clean them with approved lens paper and lens cleaner. You can calculate the **total magnifying power** by multiplying the power of the ocular with the power of the objective.

g. **Mechanical stage.** The mechanical stage includes a slide holder that looks like a clamp designed to hold the slide in place.

h. **Stage adjustment knobs.** These are knobs used to move the slide in different directions.

i. **Stage.** This is a flat surface the slide rests upon.

j. **Iris diaphragm.** This is directly underneath the stage and above the light source. This structure helps control the amount of light that passes through the slide.

k. **Light source.** This is a light bulb that sends light through the diaphragm.

l. **Base.** This is the structure that the microscope rests upon. Always carry the microscope with one hand underneath this structure.

m. **Coarse adjustment.** This knob brings objects into focus quickly.

n. **Fine adjustment.** Use this knob to sharpen the image of the object that you wish to look at.

FOCUSING THE MICROSCOPE

Obtain the letter "e" slide. When focusing the microscope you should always start with the lowest power objective. This objective has the widest field of vision and therefore allows you to see more of the slide at one time, which allows you to locate objects faster. Begin focusing using the coarse adjustment. When the object comes into focus, then you can use the fine adjustment to bring it into sharper focus. You can increase power by rotating the nosepiece to the next highest power objective. Because microscopes are **parfocal** you should only need to use the fine adjustment to bring the object into focus on higher powers. Never use the coarse adjustment knob with high power because this may break the slide.

USING THE MICROSCOPE

In order to get the most out of this lab you should be aware that microscopes have certain properties because of the way the lenses bend light. One feature is that microscopes invert images (**inversion**). In other words, right is left and up is down. Examine the letter "e" slide and notice what has happened to the image. If you have placed this slide on the stage as the "e" would normally appear, then it will be upside down and turned right to left when examined through the ocular. This isn't a problem normally; however, it can become somewhat of a hassle when you are examining something that is alive. If a microbe swims to the right of the image it is actually moving left.

Another feature is that microscopes can only focus on one plane at a time; this refers to its **depth of field** (Figure 3.3). While some objects are in sharp focus, others will be slightly out of focus. You can demonstrate this with a colored thread slide. This slide has threads of different colors stacked upon one another and therefore in different planes of focus. Obtain a colored thread slide and familiarize yourself with this property.

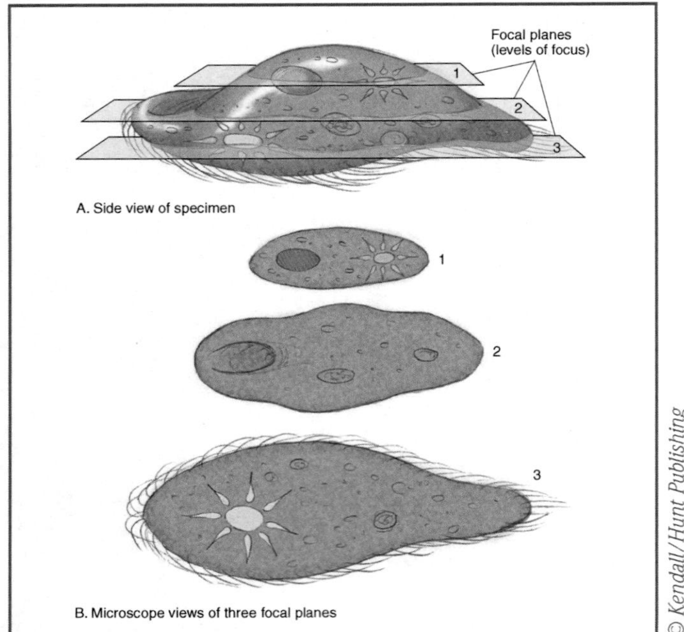

FIGURE 3.3

MAKING A WET MOUNT

You need to become familiar with making a wet mount (Figure 3.4). Use a drop of distilled water to practice this skill. When making a wet mount place a drop of liquid on the slide and then place a coverslip on top of the specimen. First place one side of the coverslip on the glass slide and then gently lower the coverslip over your specimen. This will help the suspension spread evenly across the slide and also help eliminate air bubbles. Do not press down on the coverslip as this may break it.

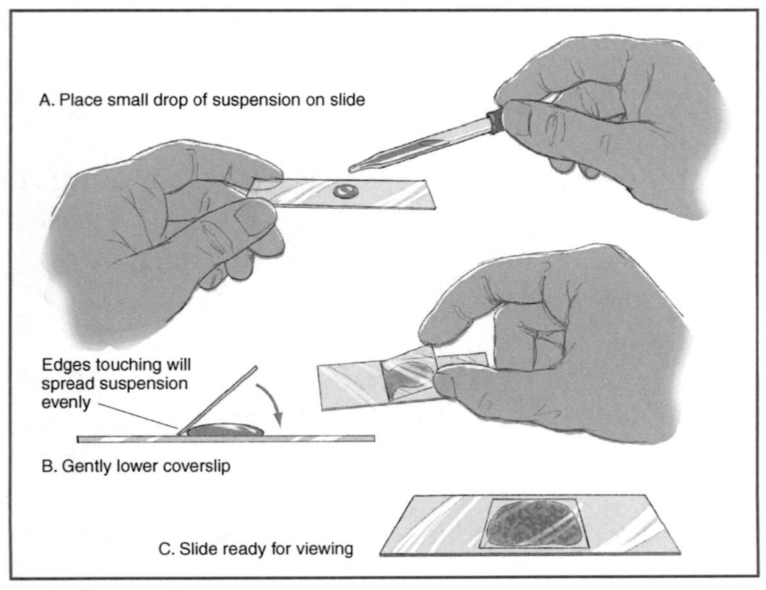

FIGURE 3.4

EXAMINATION OF CHEEK CELLS

Make a wet mount of your cheek cells for examination. Take a toothpick and GENTLY scrape the inside of your cheek. This will remove a sample of tissue you will then transfer to a clean glass slide. Once you place this material on the slide you can stain it with a drop of diluted methylene blue stain. This will help stain the nucleus of the cell to make it more visible. Draw what you see in the space provided and label the nucleus. Dispose of the toothpick as instructed.

EXAMINATION OF POND WATER

Make a wet mount of a drop of pond water provided by your instructor. Do not use methylene blue for this slide. These are living organisms, and many will be in motion so remember how the microscope inverts objects when following them. Draw what you see in the space provided.

NAME ____

EXERCISE 3

QUESTIONS

1. Define or describe the following structures:

   Rotating nosepiece

   Base

   Iris diaphragm

   Mechanical stage

2. What two parts do you use to carry the microscope?

3. What is "depth of field"?

4. What is inversion?

5. In the space below draw a microscope and label as many parts as you can.

EXERCISE 4

BASIC CHEMISTRY

INTRODUCTION

Atoms compose all matter, and living organisms are no exception. Because atoms make up all living things, it is important that you have a basic understanding of an atom's composition and behavior. That is the focus of today's lab as we explore some basic chemistry.

ATOMS, ELEMENTS, AND THE PERIODIC TABLE

Atoms are the smallest unit of an element that you can have and still have that element. For instance, you cannot have one-half of an atom of carbon and still have carbon. Atoms consist of three basic subatomic particles, **protons**, **neutrons**, and **electrons**, all of which have certain properties. Protons have a positive electrical charge, and neutrons have no charge; both of these particles make up the nucleus of an atom. Electrons have a negative charge and exist outside of the nucleus. The **periodic table** (Figure 4.1) arranges elements based on the number of protons they contain and their properties. It lists each element, its name, **chemical symbol**, atomic number and atomic weight (mass number). The **atomic number** refers to the number of protons an element has. The **atomic weight** refers to the mass of the protons plus the mass of the neutrons. You can calculate the number of neutrons an element has by subtracting the atomic number from the rounded atomic weight. For instance, hydrogen has 0 neutrons because the atomic weight rounds to 1 and the atomic number equals 1: $(1 - 1 = 0)$. All elements on the periodic table are electrically neutral because the number of positive charges (protons) is equal to the number of negative charges (electrons). Although the periodic table does not list the number of electrons in an atom, you can determine this by looking at the atomic number. Because the number of protons equals electrons the atomic number will also be the same as the number of electrons.

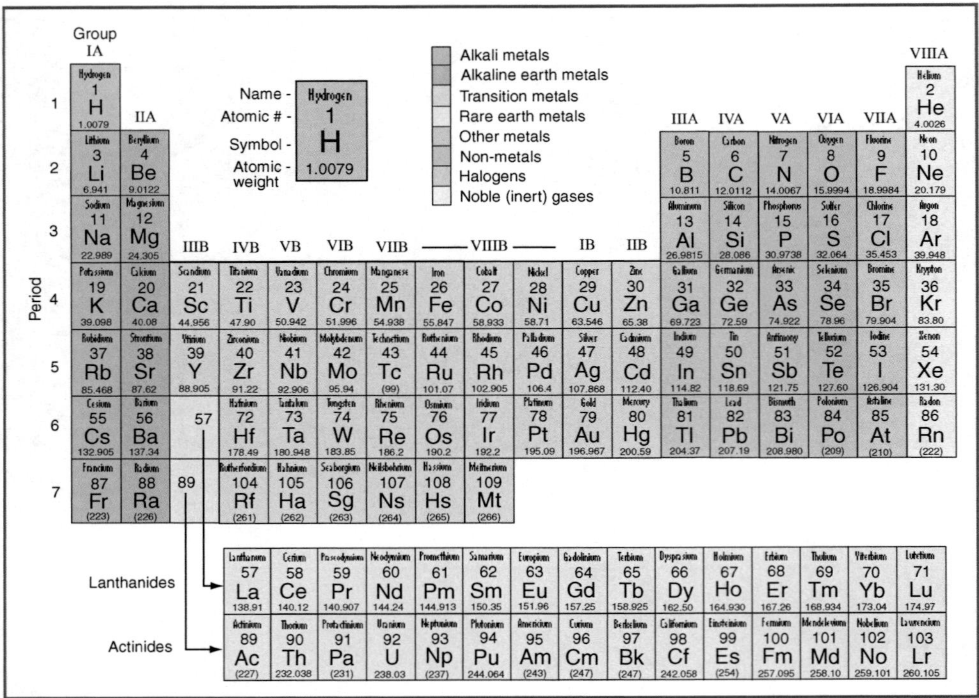

FIGURE 4.1

PERIODIC TABLE REVIEW EXERCISE

Use the periodic table to fill in the chart below.

ELEMENT	SYMBOL	ATOMIC NUMBER	ATOMIC MASS	PROTONS	NEUTRON	ELECTRONS
Carbon						
Hydrogen						
Oxygen						
Phosphorus						
Potassium						
Iodine						
Nitrogen						
Sulfur						
Calcium						
Iron						
Magnesium						
Sodium						
Chlorine						

ENERGY LEVELS

While the nucleus of an atom contains protons and neutrons, the location of electrons is more complex. Electrons are located in **orbitals**, which orbit the nucleus at varying distances called **energy levels** (Figure 4.2). The first energy level (K) has one orbital called the "s" orbital. It is spherical in shape, hence its name. The other energy levels (2nd: L, 3rd: M, 4th: N) contain an "s" orbital and a "p" orbital. The "p" orbital is dumbbell shaped and lies on 3 axis (x, y, and z). The "s" orbital can hold only two electrons, and each axis of the "p" orbital can also hold only two. The first energy level contains a maximum of two electrons because it has only one orbital, whereas, the other energy levels can hold eight each because they contain more orbitals. We can draw the energy levels and orbitals much like the orbits of the planets around a star (Figure 4.2). It is important to note that the 1s orbital, which is the "s" orbital in the first energy level (K), will fill before the others. The next to fill is the 2s orbital, and once the 2s orbital is full the 2px, 2py, and 2pz accept electrons. For the "p" orbital each axis will contain only one electron until they all contain one. Because of an electron's charge they will repel one another until they have no choice but to occupy the same orbital. Think of it like this: You would not allow someone to sit with you in your seat (orbital) as long as there are other seats (orbitals) available. Only when all seats are full will you allow someone to share your seat.

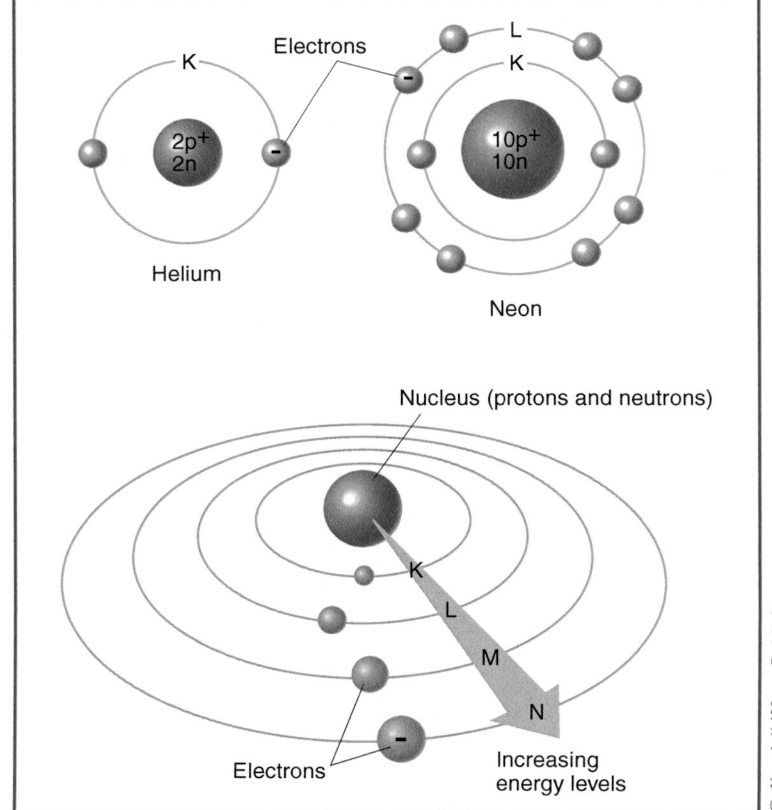

FIGURE 4.2

Take a look at the periodic table (Figure 4.1). On the left side are the **periods.** The period refers to the number of energy levels an atom has. Those elements in period 1 have one energy level (K). Those in period 2 have two energy levels (K, L). Periods 5, 6, and 7 contain not only levels K, L, M, N but energy levels O, P, and Q respectively. Look at the top of the periodic table. The vertical columms are the **groups.** Groups IA, IIA, IIIA, IVA, VA, VIA, VIIA, VIIIA will give you the number of electrons in their valence shell. A **valence shell** is the last energy level with at least one electron. We call electrons in the valence shell **valence electrons.** For instance, those in group IA have one electron in their valence shell. Those in group VIIIA have eight in their valence shell, except helium, which only has two because its valence shell is the first energy level (K). The valence shell for any given atom corresponds to the period that it is in. Remember, with the exception of hydrogen and helium, all other elements in the previously mentioned groups can contain eight electrons in their valence shells.

REVIEW EXERCISE

Use the periodic table to fill in the chart below.

ELEMENT	PERIOD	NUMBER OF ENERGY LEVELS	NAME OF VALENCE SHELL	GROUP	# VALENCE ELECTRONS	# OF ELECTRONS NEEDED TO FILL VALENCE SHELL
Carbon						
Hydrogen						
Helium						
Nitrogen						
Oxygen						
Sodium						
Calcium						
Chlorine						
Neon						
Phosphorus						
Magnesium						
Boron						
Lithium						

Use a line diagram (shown in Figure 4.2) to indicate the locations of electrons of the following elements.

a. Carbon

b. Oxygen

c. Helium

d. Nitrogen

e. Hydrogen

f. Chlorine

g. Sodium

h. Lithium

i. Argon

CHEMICAL BONDS

You might now be wondering what the point of all this is. One of the characteristics of life is a high degree of organization. However, you cannot achieve that organization without some degree of chemical complexity. Living organisms are not just a collection of elements. Life requires that molecules bond together to form more complex molecules like carbohydrates, proteins, lipids and nucleic acids, to name a few. Why do elements bond together? The answer to that question lies with their valence shell. Elements will form bonds to fill their valence shell with electrons. Elements of the main groups (those mentioned earlier) follow the **octet rule**. That is, they need eight electrons within their valence shell for it to fill completely. The exception to this is of course hydrogen and helium. Take a look at group VIIIA. These are the **noble gases** or **inert gases**. Their valence shells are already full so they will not react with other elements to make bonds. Elements within the other groups have incomplete valence shells and will therefore make bonds.

We will cover two types of bonds: ionic and covalent bonds. **Ionic bonds** form when one element strips an electron away from another. Why does this happen? Take a look at sodium and chlorine. Sodium is in group IA and therefore has one valence electron, but it needs eight to fill the third energy level. Chlorine is in group VIIA and has seven valence electrons and needs one more to fill its valence shell. Sodium has a choice; it can either gain seven more electrons or lose one. If it loses one electron, the second energy level becomes its valence shell, and it is easier to lose one than to gain seven. The opposite is true for chlorine. As a result chlorine strips one electron away from sodium to form an ionic bond. Since both atoms have a full valence shell, a stable molecule forms (NaCl). Sodium lost an electron whereas chlorine gained one electron. Sodium and chlorine are now ions (atoms with a charge. Sodium is a **cation** (positively charged ion) and has a +1 charge because it now has one more proton than electrons because it lost one electron to chlorine. This is written Na^{+1}. Chlorine is an **anion** (negatively charged ion) and has a −1 charge (Cl^{-1}).

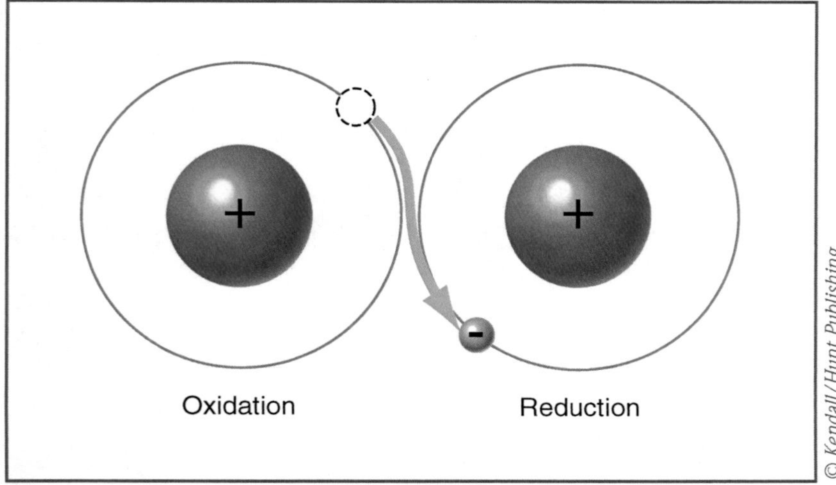

FIGURE 4.3

We call reactions where electrons are gained or lost **oxidation/reduction reactions** (Figure 4.3). The atom that lost an electron is **oxidized**, and the atom that gained an electron has been **reduced.** It is reduced because its overall charge is more negative. Remember the phrase "**LEO** the lion says **GER.**" LEO stands for "lose electrons oxidation," and GER stands for "gain electrons reduction."

Covalent bonds occur when two or more atoms share electrons. Water is a good example of this type of bond. Oxygen needs two more electrons to fill its valence shell, and hydrogen needs one. Oxygen can make bonds with two hydrogen atoms, and each hydrogen atom will make one bond with oxygen. At each bond an electron is shared to form water (H_2O). Another example is O_2, which is the free oxygen we breathe. Since oxygen requires two more electrons to fill its valence shell, they each share two, and a double bond forms. The number of electrons needed to fill its valence shell determines the number of bonds an element will make.

REVIEW

Fill in the chart below.

ELEMENT	NUMBER OF ELECTRONS GAINED OR LOST	OXIDIZED OR REDUCED
Ca^{+2}		
Li^{+1}		
F^{-1}		
P^{-3}		
Mg^{+2}		
K^{+1}		

NAME ____________________

EXERCISE 4

QUESTIONS

1. How many electrons are in the following?

   Helium

   Argon

   Potassium

   Sodium

2. What is located in the nucleus of an atom?

3. Define atomic weight (atomic mass).

4. Where are electrons located in an atom?

5. The noble gases are located in which group?

6. How many electrons can the "s" orbital hold maximum?

EXERCISE 5

CHEMICAL COMPOSITION OF CELLS

INTRODUCTION

All matter consists of atoms, composed of protons, neutrons, and electrons. However, one of the characteristics of life is a high degree of organization. As a result, very complex molecules compose living organisms. A group of molecules that are very important for life are the **biological macromolecules**. These molecules can be very complex, but they are made of smaller subunits called **monomers**. A group of monomers makes a **polymer** (macromolecule) like a group of boxcars (monomer) makes a train (polymer). For instance, **carbohydrates** are made of sugars. A **monosaccharide** (Figure 5.1) is a simple sugar, and a **disaccharide** (Figure 5.2) is made

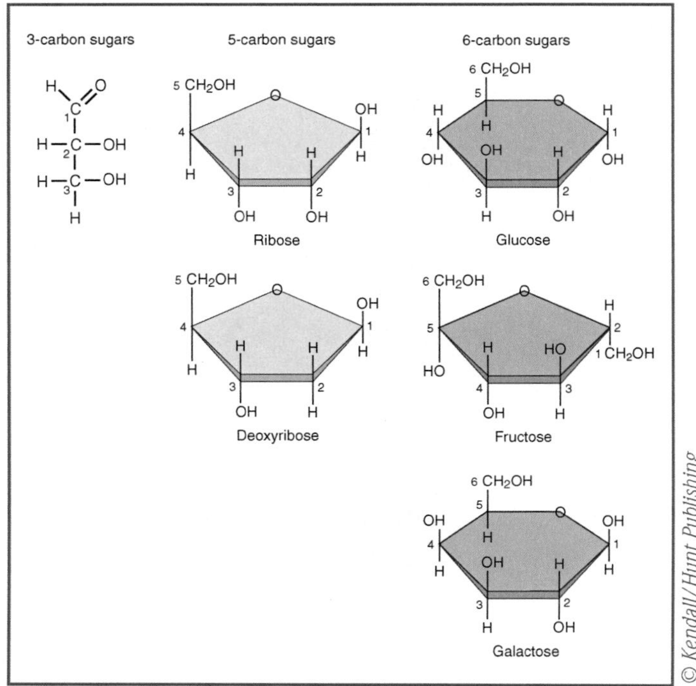

FIGURE 5.1

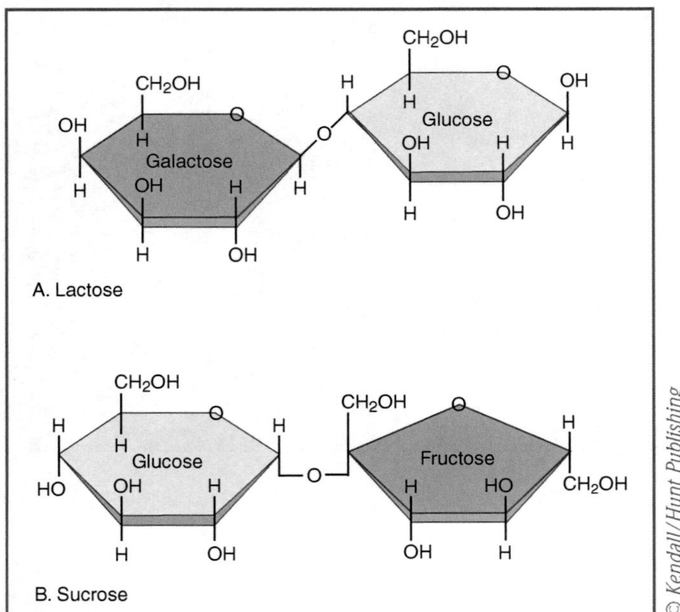

FIGURE 5.2

of two monosaccharides, whereas a **polysaccharide** is made of many monosaccharides. A **lipid** is made of a glycerol and three fatty acids. **Proteins** (Figure 5.3) are made of a chain of **amino acids** and are very complex structurally. **Nucleic acids,** which contain all of the information to make an entire organism, are actually quite simple as they consist of nucleotides that differ only in one of four **nitrogenous**

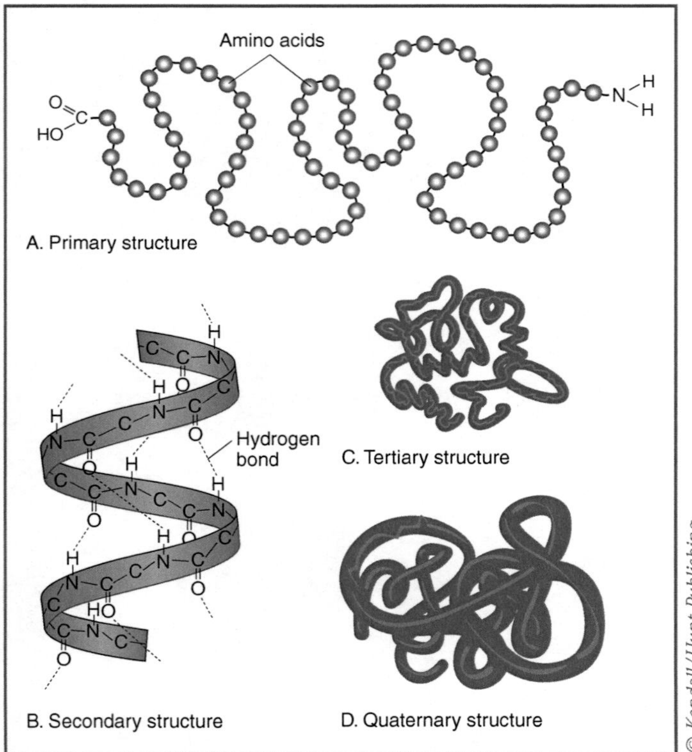

FIGURE 5.3

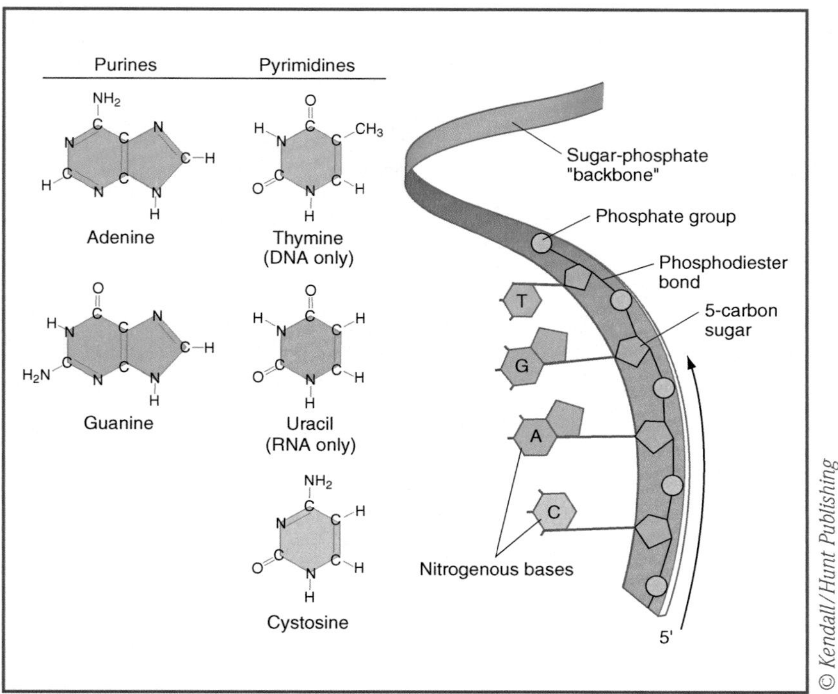

FIGURE 5.4

bases (Figure 5.4). Today you will learn how to test solutions for the presence or absence of sugars, starch, lipids, and proteins.

TESTING FOR SUGARS

Today we will test for the presence of reducing sugars using Benedict's reagent. These are sugars that in a basic solution form some aldehyde or ketone that allows the sugar to act as a reducing agent. These include fructose, glucose, lactose, and maltose. Sucrose is not a reducing sugar, and the same test cannot detect it. Benedict's solution reacts with reducing sugars by producing a color change.

Procedure

Take two test tubes and put a 1 cm mark on each test tube from the bottom. Fill test tube 1 to the mark with glucose, and put the same amount of distilled water in tube 2. Test tube 2 will serve as your control to show you what a negative test will look like. In each test tube put 5–7 drops of Benedict's reagent. The Benedict's test requires heating, so place each tube into a beaker of water over a hot plate and heat to boiling. The test tube with glucose will begin to change color, from green to yellow, orange, and red, depending on the concentration. This is a positive test.

TESTING FOR PROTEIN

You will use the Biuret's test when testing a solution for protein. Biuret solution is caustic as it contains NaOH (sodium hydroxide), and you should use caution when using this chemical.

Procedure

Take two test tubes and mark as in the previous test. Fill tube 1 to the 1 cm mark with albumin. Then fill the other to the 1 cm mark with distilled water. Test tube 2 will once again serve as your control. Then place 1–2 droppers of Biuret solution in each tube and notice the color change. Tube 2 will be blue, which indicates a negative result. The solution in tube 1 will change color immediately to a violet or purple color. Albumin is protein so obviously this will be a positive test. This test requires no heating.

TESTING FOR STARCH

Starch is a storage polysaccharide used by plants, which you can detect by using iodine (IKI).

Procedure

Take two test tubes and mark as in the previous test. Fill tube 1 to the 1 cm mark with a starch solution; then fill the other to the 1 cm mark with distilled water. Test tube 2 will once again serve as your control. Then place 1–2 droppers of IKI in each tube and notice the color change. Tube 2 will become yellow, which indicates a negative result. However, the solution in tube 1 will change color immediately to black, which indicates a positive result. Since you put starch in test tube 1, this should not come as a surprise. This test requires no heating.

TESTING FOR LIPIDS

Lipids are hydrophobic, which simply means that they do not dissolve in water. We can use this property to test for the presence of lipids using Sudan III. Sudan III is also hydrophobic and, as a result, will associate with other hydrophobic substances in solution.

Procedure

Take two test tubes and mark as in the previous test. Fill tube 1 to the 1 cm mark with vegetable oil; then fill the other to the 1 cm mark with distilled water. Test tube 2 will once again serve as your control. Then place 1–2 droppers of Sudan III in each tube and notice the color change. Tube 2 will have no color change, which indicates a negative result. However, the solution in tube 1 will change color immediately to a red or orange as the Sudan III associates with the lipid, indicating a positive result. This test requires no heating. You can also use Sudan IV for this test.

TESTING FOR UNKNOWNS

In the previous tests you knew what the results were going to be because you tested starch for the presence of starch, sugar for the presence of sugar, and so forth. However, you will not always know what your results will be when you test an unknown. Repeat all of the tests for the unknown listed below and complete the chart. For each unknown you will conduct four separate tests so you will need four tubes for each unknown.

UNKNOWN	SUGAR	STARCH	PROTEIN	LIPID
Apple juice				
Broth				
Skim milk				
Potato juice				
Instructor's unknown 1				
Instructor's unknown 2				
Instructor's unknown 3				

NAME ______________________________

EXERCISE 5

QUESTIONS

1. What chemical(s) would you use to test for the following?

   Sugar

   Lipids

   Protein

   Starch

2. What would a positive test look like for the following?

   Sugar

   Lipids

   Protein

   Starch

3. What would a negative test look like for the following?

Sugar

Lipids

Protein

Starch

4. What is the difference between monosaccharides, disaccharides, and polysaccharides?

EXERCISE 6

ENZYMES

INTRODUCTION

Life is dependent upon chemical reactions such as photosynthesis, cellular respiration, digestion, etc. We call all of the chemical reactions that occur collectively within an organism **metabolism**. Molecules called **enzymes** aid the reactions taking place. Most enzymes are proteins and have an **active site** with a specific shape designed for a specific **substrate**. The substrate is the molecule that an enzyme performs its action upon. Enzymes can perform many roles, which include breaking larger molecules down (**catabolic reactions**) or creating larger molecules (**anabolic reactions**). Regardless of what reaction they are performing, they will greatly increase the speed of the reaction. Today we will explore the role that enzymes have in chemical reactions and some of the factors that influence their activity.

DECOMPOSITION OF HYDROGEN PEROXIDE

In the following experiment you will demonstrate the effect of catalase on the breakdown of hydrogen peroxide which is summarized in the following chemical equation:

$$2H_2O_2 \rightarrow 2H_2O + O_2$$

In this reaction the enzyme catalase breaks hydrogen peroxide into water and oxygen. Hydrogen peroxide is the substrate that catalase performs its action upon. You will need three test tubes to perform this experiment.

Tube 1: Fill tube 1 with approximately 30 drops of hydrogen peroxide and 5 drops of catalase and mix the solution. Measure the amount of bubbles produced (if any) from the bottom of the bubble column to the top with a ruler in mm. Record your results here.

Height of bubble column ______________________

Tube 2: Fill tube 1 with approximately 30 drops of hydrogen peroxide and 5 drops of water and mix the solution. Measure the amount of bubbles produced (if any) from the bottom of the bubble column to the top with a ruler in mm. Record your results here.

Height of bubble column ______________________

Tube 3: Fill tube 1 with approximately 30 drops of sucrose and 5 drops of catalase and mix the solution. Measure the amount of bubbles produced (if any) from the bottom of the bubble column to the top with a ruler in mm. Record your results here.

Height of bubble column ______________________

Answer the following questions based on the results of this experiment.

1. Explain why each of the tubes showed different amounts of bubbles produced. In other words explain your results.

2. What were the bubbles made of (what gas)?

3. Was there a control tube for this experiment? Why is this important?

4. If you had replicated this experiment do you think you would see different results? Did your lab group see different results from other lab groups within the class? Why or why not?

EFFECTS OF TEMPERATURE

According to the **kinetic theory** molecular motion increases as the temperature of a substance increases. Because the molecules in a reaction will move faster if the temperature is higher, this should speed up a reaction. You will test this principle in the following experiment. You will need four test tubes for this experiment.

Tube 1: Mark the tube at the 1 cm and the 4 cm level. Fill the tube to the 1 cm mark with room temperature catalase and then to the 4 cm mark with hydrogen peroxide. Mix the solution and measure the amount of bubbles in mm.

Height of bubble column ______________________

Tube 2: Mark the tube at the 1 cm and the 4 cm level. Fill the tube to the 1 cm mark with catalase and put it on ice for ten minutes. Next, fill to the 4 cm mark with hydrogen peroxide chilled for ten minutes. Mix the solution and measure the amount of bubbles in mm.

Height of bubble column ______________________

Tube 3: Mark the tube at the 1 cm and the 4 cm level. Fill the tube to the 1 cm mark with catalase and place it in a water bath at 37°C for ten minutes. Next fill to the 4 cm mark with hydrogen peroxide. Mix the solution and measure the amount of bubbles in mm.

Height of bubble column ______________________

Tube 4: Mark the tube at the 1 cm and the 4 cm level. Fill the tube to the 1 cm mark with catalase. Place the tube in a beaker of water and bring to a boil. Next fill to the 4 cm mark with hydrogen peroxide. Mix the solution and measure the amount of bubbles in mm.

Height of bubble column ______________________

Answer the following questions based on the results of this experiment.

1. Did the tubes show different amounts of bubbles produced? Why?

2. Is there anything significant about 37°C?

3. Was there a control tube for this experiment? Why is this important?

4. If you had replicated this experiment do you think you would have seen different results? Did your lab group see different results from other lab groups within the class? Why or why not?

EFFECTS OF PH

Many biological processes have an optimal pH for peak efficiency. Enzymes are no exception to this. This experiment will test the effects of pH on enzyme activity. You will need three test tubes, each marked at the 1, 3, and 5 cm level.

Tube 1: Fill to the 1 cm level with catalase. Fill to the 3 cm mark with distilled water and to the 5 cm mark with hydrogen peroxide. Mix the solution and record results below. Use litmus paper to measure the pH of the solution.

Height of bubble column ______________________________

pH of solution ______________________________

Tube 2: Fill to the 1 cm level with catalase. Fill to the 3 cm mark with water mixed with HCl adjusted to a pH of 2. **Do this before pouring in the test tube**. Next fill to the 5 cm mark with hydrogen peroxide. Mix the solution and record results below.

Height of bubble column ______________________________

pH of solution ______________________________

Tube 3: Fill to the 1 cm level with catalase. Fill to the 3 cm mark with water mixed with NaOH adjusted to a pH of 12. **Do this before pouring in the test tube**. Next fill to the 5 cm mark with hydrogen peroxide. Mix solution and record results below.

Height of bubble column ______________________________

pH of solution ______________________________

Answer the following questions based on the results of this experiment.

1. Did the tubes show different amounts of bubble produced? Why?

2. Blood has a pH of 7.5. What would happen if the pH of blood was variable?

3. Was there a control tube for this experiment and why is this important?

4. If you had replicated this experiment do you think you would have seen different results? Did your lab group see different results from other lab groups within the class? Why or why not?

NAME ______________________________

EXERCISE 6

QUESTIONS

1. What are enzymes used for?

2. What is a substrate?

3. What is a catabolic reaction?

4. What is an anabolic reaction?

5. What is (are) the products and the reactant of the chemical reaction studied today?

EXERCISE 7

CELLULAR RESPIRATION

INTRODUCTION

All living things require energy, and the energy currency of the cell is adenosine triphosphate (**ATP**). It should come as no surprise that a complex biochemical process called **cellular respiration** converts the food we consume to this molecule. Aerobic cellular respiration occurs in the presence of oxygen (Figure 7.1) and has the general chemical equation below:

$$C_6H_{12}O_6 + 6O_2 \rightarrow 6CO_2 + 6H_2O + ATP$$

This process includes **glycolysis**, the **Kreb's cycle**, the **electron transport chain** and **chemiosmosis.** All of these occur in the **mitochondria** except glycolysis, which occurs in the cytoplasm of the cell. During this process **glucose** ($C_6H_{12}O_6$)

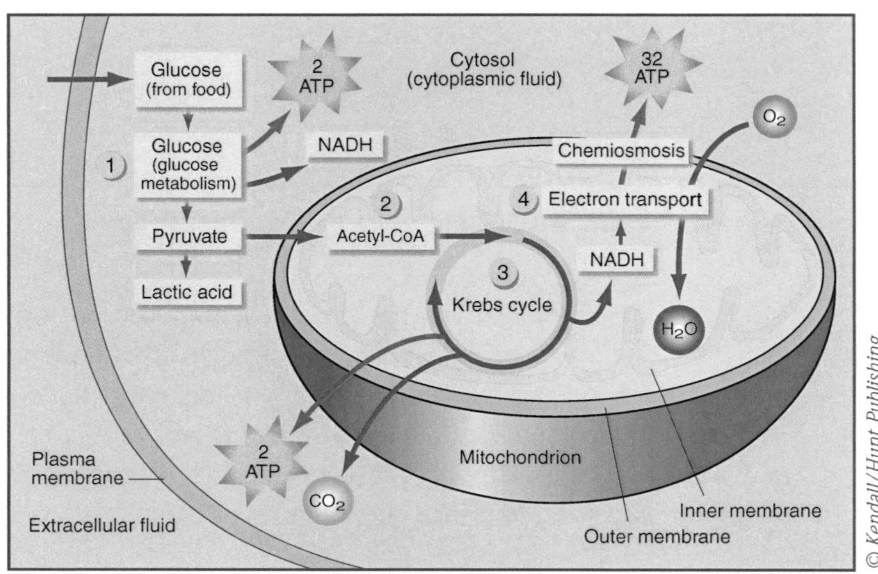

FIGURE 7.1

and oxygen become converted into carbon dioxide, water, and ATP. Anaerobic respiration does not include oxygen, the Kreb's cycle or the electron transport chain and chemiosmosis. It also does not utilize the mitochondria. The steps in this process are glycolysis and either lactic acid fermentation or alcohol fermentation. We will explore these reactions in today's experiment.

LACTIC ACID FERMENTATION

Lactic acid fermentation occurs in animal cells when the supply of oxygen has been depleted. When this happens glucose is broken down via glycolysis, but instead of **pyruvate** (made during glycolysis) entering the mitochondria, as would normally occur, fermentation begins and lactic acid is produced. This is a waste product and causes a burning sensation in the muscles. We can easily demonstrate this process. Take a volunteer and have him stand upright with arms outstretched parallel to the floor and palms upright. Place textbooks on the palms of the hands until the subject can just barely keep his arms level and wait. At some point the amount of oxygen used to make ATP will run out and fermentation will begin. When this happens the muscles begin to burn and get weaker.

1. How long did it take for fermentation to begin?

2. What caused the burning sensation in the muscles?

3. Why do you get weaker during this type of respiration?

ALCOHOL FERMENTATION

Alcohol fermentation occurs when alcohol is produced instead of lactic acid. Carbon dioxide is also produced. Yeast is a type of fungus that undergoes this type of respiration. The alcohol produced is ethanol, used in alcoholic drinks we use today. Bakers use the carbon dioxide to make breads and cakes rise. We will demonstrate this process using yeast and several possible food sources. You will need four fermentation flasks and four beakers (60 ml).

Flask 1: Pour 50 ml (you may need to vary this amount depending on the size of the fermentation flask) of water in a beaker and thoroughly mix with 1 g of yeast. Pour this solution into the fermentation flask and gently tip backward so the top of the flask is completely filled and there are no air bubbles.

Flask 2: Pour 50 ml (you may need to vary this amount depending on the size of the fermentation flask) of glucose in a beaker and thoroughly mix with 1 g of yeast. Pour this solution into the fermentation flask and gently tip backward so the top of the flask in completely filled and there are no air bubbles.

Flask 3: Pour 50 ml (you may need to vary this amount depending on the size of the fermentation flask) of fructose in a beaker and thoroughly mix with 1 g of yeast. Pour this solution into the fermentation flask and gently tip backward so the top of the flask in completely filled and there are no air bubbles.

Flask 4: Pour 50 ml (you may need to vary this amount depending on the size of the fermentation flask) of sucrose in a beaker and thoroughly mix with 1 g of yeast. Pour this solution into the fermentation flask and gently tip backward so the top of the flask in completely filled and there are no air bubbles.

Incubate all flasks in an incubator or water bath at 37°C. Measure the amount of gas that collects in the top of the flask (if any) using a ruler and record your results:

Flask 1: ______________________________

Flask 2: ______________________________

Flask 3: ______________________________

Flask 4: ______________________________

1. Did the amount of gas produced between flasks differ? Why or why not?

2. What was the gas produced?

3. Were there any organelles involved in this process? What are they, if any?

4. Was there a control in this experiment, and why is this important?

AEROBIC RESPIRATION

During aerobic respiration oxygen is consumed and carbon dioxide is produced. We can measure the production of carbon dioxide by using phenol red. Phenol red is a pH indicator that turns yellow in acidic solutions. When carbon dioxide dissolves in water the pH of the solution lowers because hydrogen ions are produced that make the solution more acidic. We will take advantage of this to measure the presence or absence of aerobic respiration in three groups of peas. You will need four test tubes and three groups of peas (dead, dormant, and germinating).

Tube 1: Place 4–5 drops of phenol red solution in the bottom of the tube. Take a soda straw and cut it into pieces. Place the straw pieces in the bottom of the tube so that their tops are just above the solution. Make sure that the solution does not change color when you place the straws in. If it does you must start over.

Tube 2: Place 4–5 drops of phenol red solution in the bottom of the tube. Take a soda straw and cut it into pieces. Place the straw pieces in the bottom of the tube so that their tops are just above the solution. Make sure that the solution does not change color when you place the straws in. If it does you must start over. Place 4–5 germinating seeds on top of the straws, but do not allow the seeds to touch the solution.

Tube 3: Place 4–5 drops of phenol red solution in the bottom of the tube. Take a soda straw and cut it into pieces. Place the straw pieces in the bottom of the tube so that their tops are just above the solution. Make sure that the solution does not change color when you place the straws in. If it does you must start over. Place 4–5 dormant seeds on top of the straws, but do not allow the seeds to touch the solution.

Tube 4: Place 4–5 drops of phenol red solution in the bottom of the tube. Take a soda straw and cut it into pieces. Place the straw pieces in the bottom of the tube so that their tops are just above the solution. Make sure that the solution does not change color when you place the straws in. If it does you must start over. Place 4–5 dead seeds on top of the straws, but do not allow the seeds to touch the solution.

Observe the test tubes for one hour noting any color change in the solution every ten minutes. This is a subjective observation as there is no way to quantify the amount of color change other than what tube you believe shows the greatest amount of it. Record the tubes from greatest to least color change below. It is possible that you may require this reaction to run overnight.

1. What was the chemical reaction of the process you just observed?

2. Was there a difference in the amount of color change between the tubes? Why?

3. Were there any organelles involved in this process? What are they, if any?

4. Was there a control in this experiment? Why is this important?

NAME ___________________________________

EXERCISE 7

QUESTIONS

1. What happens to pyruvate during lactic acid fermentation?

2. What type of alcohol is produced when yeast performs fermentation?

3. Which steps of aerobic cellular respiration occur in the mitochondria?

4. What is the chemical equation for aerobic cellular respiration?

5. What are the reactants and products of aerobic cellular respiration?

6. What are the differences between aerobic and anaerobic respiration?

EXERCISE 8

PHOTOSYNTHESIS

INTRODUCTION

Photosynthesis is a cellular process where plants and some protists convert light energy into chemical energy. Following is the generalized chemical equation:

$$6CO_2 + 12H_2O \rightarrow C_6H_{12}O_6 + 6H_2O + 6O_2$$

During this process carbon dioxide and water convert into glucose, water, and oxygen. Oxygen is a waste product, and photosynthesis is responsible for all the free oxygen in the atmosphere that we breathe. This reaction is dependent upon light; Figure 8.1 shows the **light dependent reactions**. During this process light energy is captured by **chlorophyll**, which is located in the **chloroplast** of plants (Figure 8.2), and produces ATP and NADPH, used during the light independent reaction (**Calvin cycle**) of photosynthesis to make glucose. We will explore this process in today's lab.

CARBON DIOXIDE UPTAKE DURING PHOTOSYNTHESIS

During the Calvin cycle carbon dioxide enters the plant through the leaf stomata. Carbon fixation occurs in this process, and the carbons are converted into glucose. We can monitor the uptake of carbon dioxide by using phenol red. Phenol red is an acid indicator that turns yellow in an acid and red in a base. As the plant takes up carbon dioxide the number of hydrogen ions will decease and the solution will become more basic. As a result, the solution's color will change to red.

Take a generous piece of the aquatic plant *Elodea* and place it in a test tube with a diluted solution of phenol red. Next take a straw and gently blow into the solution to decrease its pH. This causes carbon dioxide, provided by your breath. It dissolves in water and increases hydrogen ion concentration. This process lowers

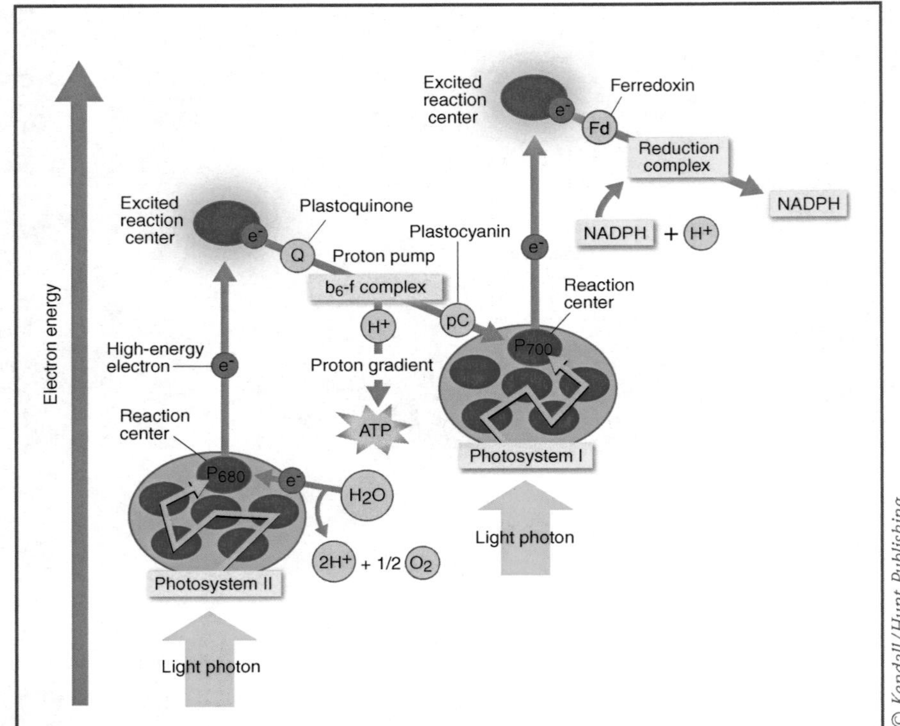

FIGURE 8.1

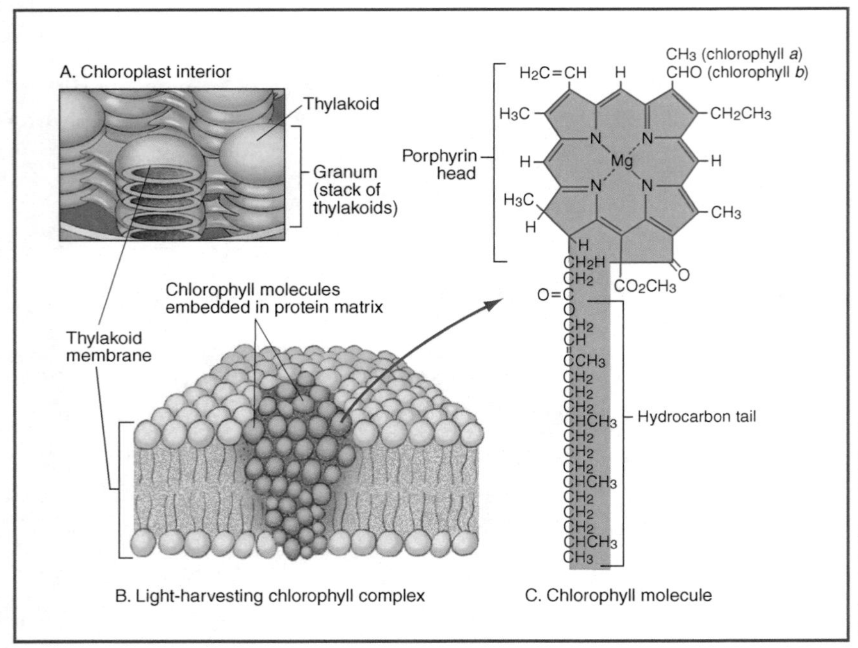

FIGURE 8.2

the pH of the solution and turns phenol red to yellow. As soon as the solution turns yellow stop blowing into the solution. Place the tube 1 m from a light source and monitor it every ten minutes for one hour, watching for a color change in the phenol red solution.

1. What does the color change indicate?

2. During what stage of photosynthesis does carbon dioxide uptake take place?

PLANT PIGMENTS

Light is, of course, a requirement for photosynthesis. However, white light is a combination of various wavelengths that we can demonstrate by exposing white light to a prism. Plants have a variety of pigments designed to absorb certain wavelengths. The reason you see a color is because the pigment did not absorb that wavelength but reflected it back to your eyes. For instance, green pigment does not absorb green wavelengths, and therefore you see the plant as green. We can examine the pigments a plant makes through paper chromatography.

1. Obtain one test tube, a stopper with a hook, and a piece of chromatography paper touching only the top of the paper.

2. Attach the piece of chromatography paper to the hook and hang it in the test tube to test for fit. The bottom of the paper should just touch the bottom of the tube without bending.

3. Place a small dot with a pencil 2 cm from the bottom of the paper.

4. Mark the bottom of the test tube 1 cm from the bottom.

5. Obtain some spinach leaves and crush them with a mortar and pestle. Add small amounts of acetone to help remove pigments.

6. Once a dark green liquid has formed, place a small drop of the pigment extract on the dot made by the pencil using a capillary tube as shown in Figure 8.3.

7. You need to obtain a small green dot of plant extract so you may need to repeat step 6 several times, allowing the extract to dry each time.

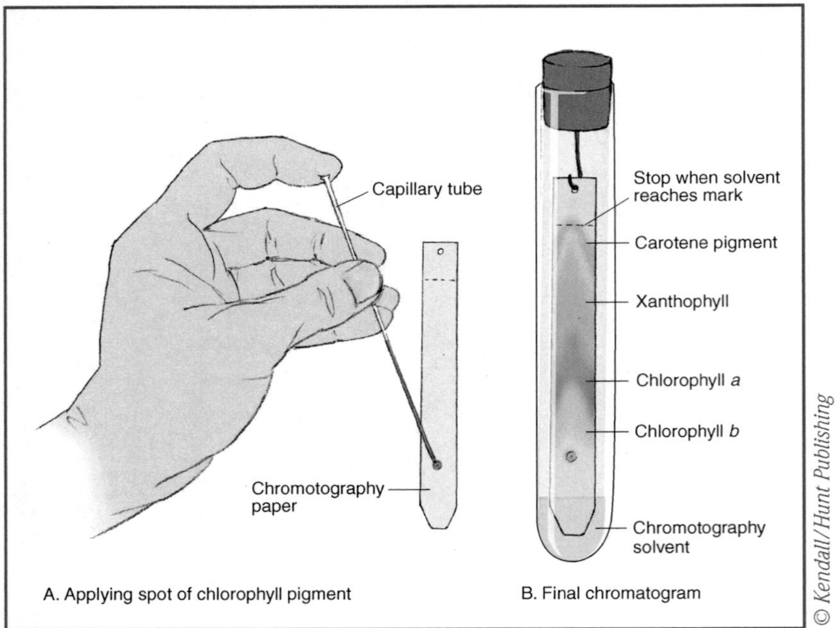

FIGURE 8.3

8. Once completed, pour a small amount of chromatography solution into the tube filling to the 1 cm mark. Do this in a fume hood.

9. With the chromatography paper attached to the hook on the cork, place in the tube with the paper inserted in the solution. Make sure not to submerge the dot in the solution.

10. Place level on a test tube rack and do not disturb.

11. The chromatography solution will move up the paper, separating the pigments. Check every few minutes and remove the stopper with the paper attached when the solution reaches 1 cm from the top. Allow to dry. Your chromatogram should look similar to Figure 8.3.

12. Identify all pigments separated. Place a small dot in the center of each band. Also, mark the furthest movement of the solution with a pencil.

13. You can calculate the R_f (ratio-factor) value of each pigment by measuring (in mm) the distance from the initial spot to the spot in the center of the band divided by the distance moved by the solvent.

    R_f = *Distance moved by pigment*
    Distance moved by solvent front
    measuring from the original dot.

Record your results in the chart below:

PIGMENT	DISTANCE (MM)	R_f VALUE

Answer the following questions.

1. Which wavelengths of light does each pigment reflect?

2. Which pigments had the greatest/lowest R_f values?

3. Why do you think plants have so many types of pigments? (Why is it important/ adaptive)?

EFFECTS OF LIGHT

During this experiment we will demonstrate the effects of filtering out various wavelengths of light on photosynthesis by measuring the amount of oxygen produced. You will need a total of six test tubes with stoppers and pipettes for this experiment.

Obtain six test tubes and fill all six with 3% sodium bicarbonate. Place a generous quantity of *Elodea* in each.

Tube 1: Cover with aluminum foil to block all light.

Tube 2: Do nothing else to it.

Tubes 3–6: Place 2–3 drops of food coloring in each tube as shown in Figure 8.4 using red, green, yellow and blue. You may utilize other colors if those are not available, but you must use green.

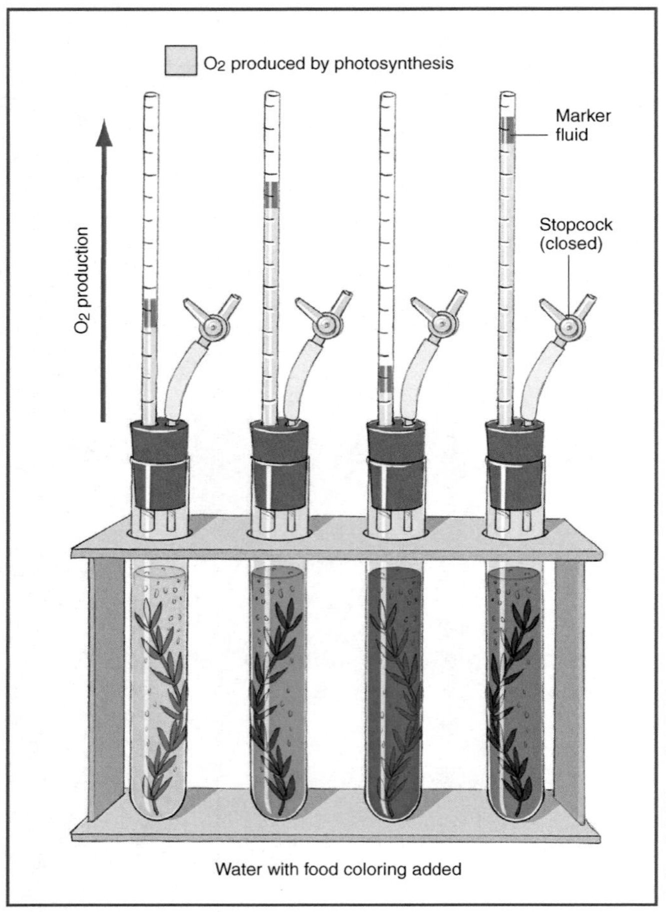

FIGURE 8.4

Stopper all tubes (1–6) so that fluid moves up into the tube. If no marker fluid is available pay close attention to the level of fluid. Allow ten minutes for the *Elodea* to acclimate to its environment. After the ten minutes place all tubes 1 m from a light source and mark the level of fluid in each tube with a grease pencil. Monitor the fluid level in the tubes every ten minutes for one hour.

1. How far did the fluid level rise or fall in each tube?

2. Compare the amount of fluid level movement for each tube and explain your results.

STARCH PRODUCTION IN PLANTS

The whole purpose of photosynthesis is to make food for the plant. The glucose made will eventually become ATP via cellular respiration. However, plants will store glucose in the form of starch. Starch is a polysaccharide used to store energy. We can measure the production of this molecule by examining leaves that have had exposure to light compared to those that have had none.

For this experiment we will use *Geranium* leaves, which foil has partially covered for one week to prevent that part of the leaf from performing photosynthesis. Take a leaf provided to you and draw a sketch of it indicating in the space below the area of the leaf that foil had covered.

Next boil the leaf in water to remove any water-soluble pigments. After boiling in water, boil the leaf in alcohol. This removes the chlorophyll from the plant.

Alcohol is extremely flammable so use a hot plate and not a Bunsen burner.

Once this is accomplished, gently dry the plant. Next pour a solution of IKI on the leaf. Draw another sketch of the plant in the space below indicating the areas that stained dark after the addition of IKI.

1. Do the dark areas stained by IKI match the areas exposed to light or those kept from light? Why?

2. What substance did the IKI test for?

3. What organelle contains starch?

NAME ____

EXERCISE 8

QUESTIONS

1. What would happen to a plant if it were exposed only to green light?

2. What is the chemical equation for photosynthesis?

3. What are the reactants and products of photosynthesis?

4. NADPH and ATP are used during what step of photosynthesis?

5. What molecule does a plant use to store energy?

6. What organelle does photosynthesis occur in?

EXERCISE 9

CELL STRUCTURE

INTRODUCTION

According to the cell theory (Schleiden and Schwann), at least one cell composes each living thing. Additionally the cell theory states that cells are the basic unit of life and that all cells come from other cells. Cells come in two general forms and are classified based on the presence or absence of certain structures. These structures are called **organelles**, which literally means "little organ." **Prokaryotic cells** (Figure 9.1) are the simplest form of life and are not as complex as **eukaryotic cells** (Figure 9.2). The major difference between these two types of cells is their internal organization. Eukaryotic cells are more complex with membrane-bound organelles, such as a true nucleus. In today's lab you will learn the organelles and the basic differences between prokaryotic and eukaryotic cells.

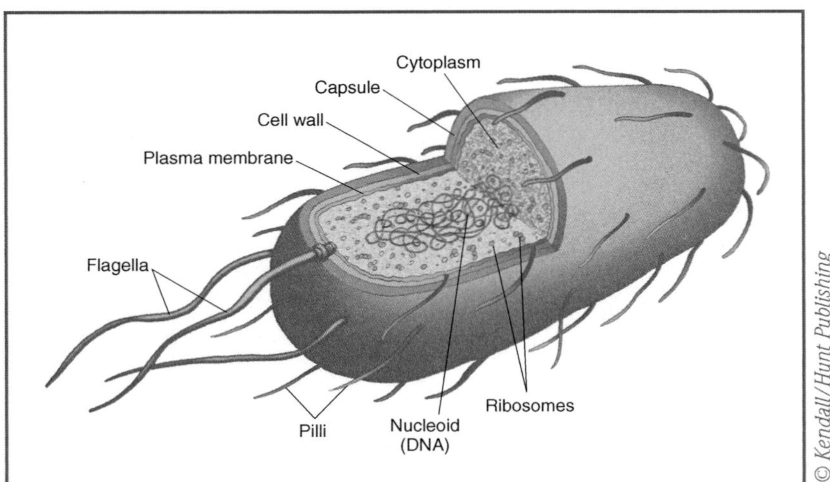

FIGURE 9.1

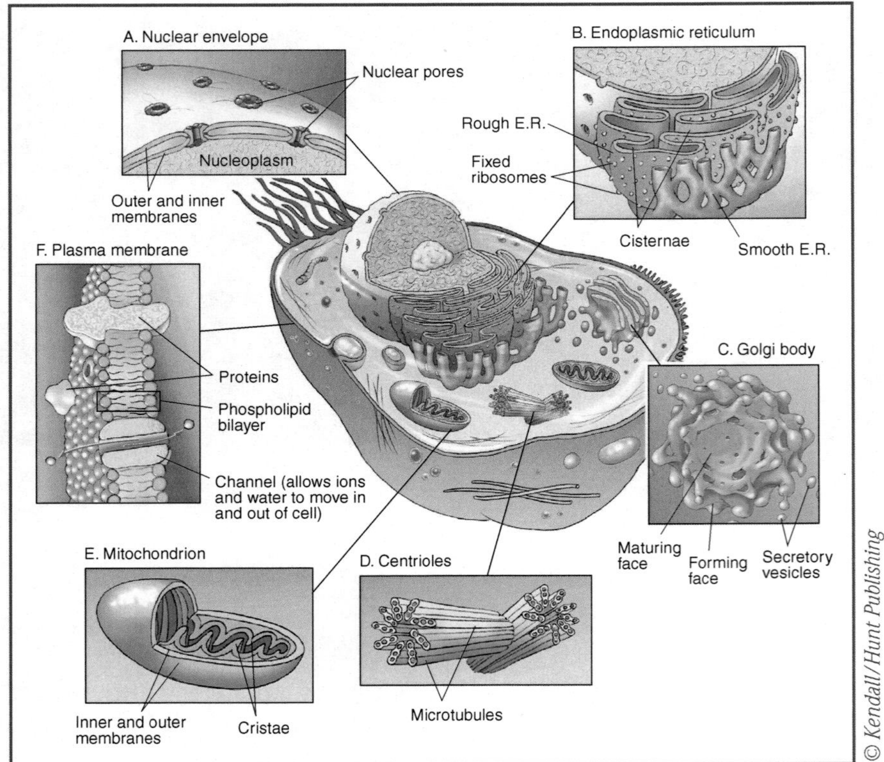

FIGURE 9.2

ORGANELLES

Organelle	*Function*
Cell wall	Support and protection for the cell
Plasma membrane	Forms the boundary of the cell, used to regulate the passage of materials in and out of the cell
Nucleus	This separates the genetic material from the rest of the cell and is the site of DNA replication
Nucleolus	Located in the nucleus, this is the site of ribosome formation.
Ribosome	Protein synthesis
Endoplasmic Reticulum	
Rough ER	Contains ribosomes so it looks rough; used in the synthesis of proteins
Smooth ER	Contains no ribosomes; used to synthesize lipids, to detoxify drugs, and to store calcium ions
Golgi apparatus	Used to process, package and ship proteins and lipids to other locations

Vacuoles and vesicles	Used for storage; vacuoles are typically larger than vesicles
Lysosome	A vesicle containing digestive enzymes
Mitochondria	Site of aerobic cellular respiration
Chloroplast	Used for photosynthesis
Peroxisome	A vesicle used for metabolic tasks such as the breaking down of fatty acids
Plastids	Can contain pigment such as in the chloroplast; some contain food such as starch
Cytoskeleton	The internal skeleton of the cell used for support and movement; made of microtubules and intermediate filaments
Microtubules	Used for support; also the structural component of other organelles
Microfilaments	Used for structural support
Intermediate filaments	Used for support
Cilia and Flagella	Made of microtubules (9+2 array); used for movement
Centrioles	Made of microtubules; used during cellular division

ANIMAL CELLS

Figure 9.3 shows a generalized animal cell. Examine this figure and label the following parts: centriole, chromatin, cilia, cytoplasm, cytoskeleton, exocytotic vesicle, fixed ribosome, free ribosomes, Golgi apparatus, lysosome, microfilaments, microtubules, microvilli, mitochondria, nuclear envelope, nuclear pore, nucleolus, nucleoplasm, nucleus, peroxisome, plasma membrane, polyribosome, rough endoplasmic reticulum, secretory vesicle, smooth endoplasmic reticulum.

PLANT CELLS

Figure 9.4 shows a generalized plant cell. Examine this figure and label the following parts: cell wall, cell walls of adjacent cells, central vacuole, chloroplast, chromatin, cytoplasm, Golgi apparatus, microtubules, mitochondria, nuclear envelope, nuclear pore, nucleolus, nucleus, plasma membrane, rough endoplasmic reticulum, smooth endoplasmic reticulum, starch plastid.

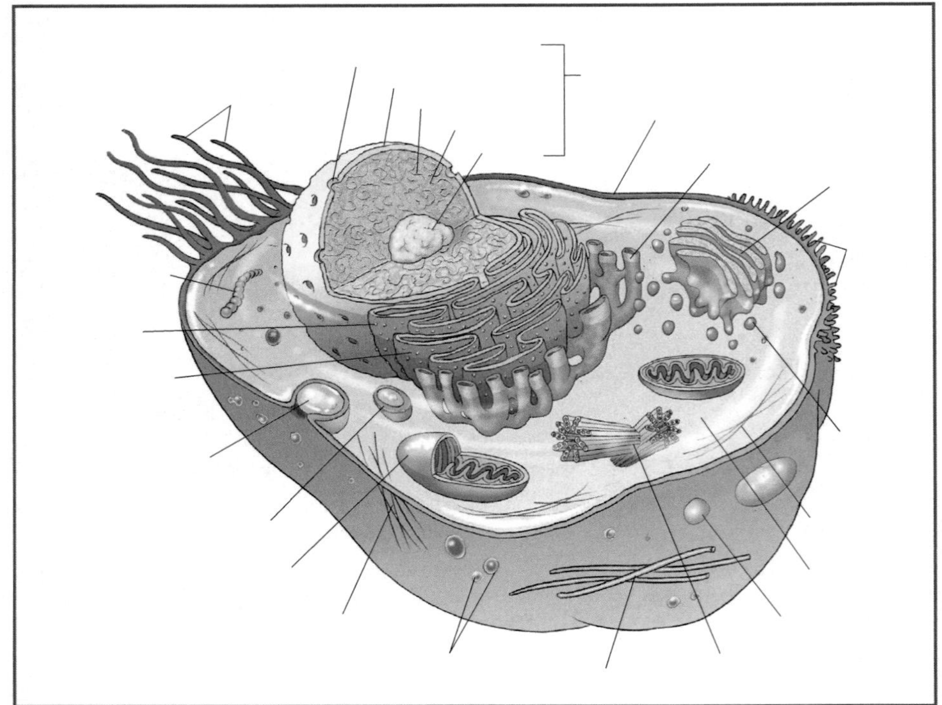

FIGURE 9.3

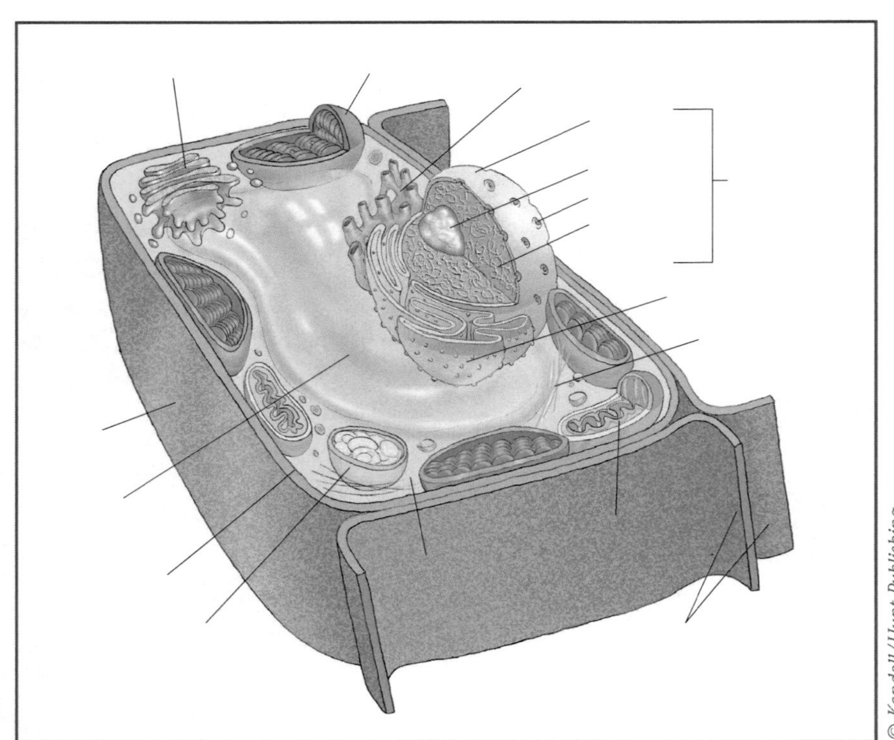

FIGURE 9.4

COMPARISON OF PROKARYOTIC AND EUKARYOTIC CELLS

Complete the chart below and indicate whether the listed organelles are found in prokaryotic and eukaryotic cells. Use the lab book, textbook, and class notes as references.

ORGANELLE	PROKARYOTIC (YES/NO)	EUKARYOTIC (YES/NO)
Cell wall		
Nucleus		
Nucleolus		
Endoplasmic reticulum		
Ribosomes		
Golgi apparatus		
Lysosomes		
Mitochondria		
Chloroplast		
Plasma membrane		
Cilia/flagella		

MICROSCOPIC EXAMINATION OF ANIMAL CELLS

Examine prepared slides of animal tissue such as epithelial tissue, nervous tissue, bone, cartilage, blood, and adipose tissue. Draw what you see in the space below and label all identifiable structures.

MICROSCOPIC EXAMINATION OF PLANT CELLS

Take a leaf from the aquatic plant of the genus *Elodea.* Do not remove an entire stem, but pull a leaf off as close to the stem as possible. Make a wet mount using distilled water. Place on the stage of the microscope and begin focusing while observing on the lowest power objective. Increase to the next highest power once in focus if desired.

The green structures that are evident are chloroplasts. They are green because they contain **chlorophyll.** Chlorophyll is the pigment used by plants for photosynthesis. The rigid outer structure of each cell is the cell wall. You also may notice that there is an apparent hollow space in many of the cells. This is the large central vacuole found in plant cells and not animal cells. You will find the chloroplasts in these cells located around the vacuole. In some cells there may appear that there is no vacuole because the chloroplasts are located everywhere. This is because these cells are three-dimensional, and the chloroplasts are on top of the vacuole. You also may notice movement within the cell as the chloroplasts move around. This is called **cytoplasmic streaming.** The cytoplasm is the liquid potion of the cell and will flow throughout the cell, which causes the chloroplasts to move around.

Use the space below to draw what you see and label the cell wall, chloroplasts, large central vacuole, and the plasma membrane.

NAME ______________________________

EXERCISE 9

QUESTIONS

1. Define the cell theory.

2. What are the differences between eukaryotic and prokaryotic cells?

3. What are the differences between plant and animal cells?

4. Give the function of the following organelles.

   Nucleolus

   Smooth endoplasmic reticulum

   Plastids

Centrioles

Peroxisomes

Mitochomdria

EXERCISE 10

CELL MEMBRANE PERMEABILITY

INTRODUCTION

The **cell** is the basic unit of structure and function of life. In 1838–39 Schleiden and Schwann stated the **cell theory**: all living organisms are composed of cells. Cells vary in function, shape, size, and structure. Within a cell there may be several types of organelles, each having a typical structure and function of its own.

An example of a cell organelle is the **cell membrane** (plasma membrane). The purpose of this exercise is to introduce the student to some of the phenomena which help explain the function of the cell membrane.

The cell membrane and the membranes of cell organelles appear as a bilayer when viewed with an electron microscope. This bilayer is a double layer of phospholipid molecules with proteins associated in various ways. Some membranes also contain sterols. The phospholipid bilayer is considered to be the structural component of the membrane, with the proteins being the functional component. Proteins have been shown to be very dynamic in terms of position in or on the membrane. The phospholipid bilayer is highly **hydrophobic** (non-water soluble) and the proteins are usually **hydrophilic** (water soluble), accounting for some characteristics of the membrane. Many of the proteins have a role in transport of molecules across the membrane. Proteins in this role are selective as to what will be transported. Transport proteins function in the process known as active transport which requires cellular energy in the form of ATP. Active transport is the only membrane transport mechanism that will concentrate molecules.

Nonpolar hydrophobic molecules such as lipids and the lipid-soluble gases, oxygen and carbon dioxide, diffuse readily across cell membranes. Cell and organelle membranes fit the **fluid mosaic model** for membrane structure and function (see your text). The percentages of phospholipid, protein, and sterol molecules in cell membranes vary with the cell type. Protective cells (skin) have a higher percentage of lipids while the lining cells of the intestine which absorb food molecules have more protein in their membranes.

The cell membrane is a characteristic of life since it is found in all cells. Its role is to isolate the cell from the external environment and to regulate the passage of materials into and out of the cell. The cell membrane and the cell wall are completely different cell parts. Cells which have a cell wall (Domains Bacteria & Archaea, Kingdoms Fungi & Plantae) secrete the cell wall building blocks through the cell membrane. The cell wall is always outside the cell membrane and is not a part of the living material. Animal cells do not have cell walls. Cell walls help maintain cell shape, provide skeletal strength, and resist osmotic damage from excess water entering the cell.

NOTE: Students should conduct activities 1 and 2 while working in groups. After completing the initial set-up for activities 1 and 2, each student should individually begin activity 3. Complete activity 3 while the results from activities 1 and 2 are being collected and recorded.

ACTIVITY 1

Diffusion

Diffusion is the movement of molecules from areas of higher concentration to areas of lower concentration. This is due to molecular motion and is directly influenced by temperature. Cells are separated from their environment by a selectively permeable membrane.

The cell membrane will permit water molecules and substances such as oxygen, carbon dioxide, and lipid soluble materials to diffuse into and out of the cell. The movement of these gases and other molecules through the cell membrane is essential to the living cell.

Most hydrophilic organic molecules such as carbohydrates, amino acids, and proteins do not freely diffuse through the membrane due to their large size and electrochemical properties. These organic substances produced in one part of the cell and used in another part of the same cell may diffuse from the area of production to the area of use. Diffusion is a slow process that does not require cell energy; however, it is speeded up by metabolic activities and cytoplasmic movements.

To demonstrate diffusion in water:

1. Place approximately 3 cm of water into a test tube and stand it in a test tube rack,
2. Drop a crystal of potassium permanganate in the tube,
3. Place the test tube rack on your work table and do not move it during the laboratory period—observe periodically to see the slow diffusion of potassium permanganate.

ACTIVITY 2

Osmosis (in a Cell Model)

Water and some small ions move freely through a selectively permeable membrane and this movement of water (a type of diffusion) is known as **osmosis.** The cytoplasm of a cell consists of water molecules (averages 70%) and many other substances dissolved in the water (**solutes**). The net movement of water across the plasma membrane is affected by the amount of solute dissolved in it. Osmosis is an essential process in the living cell. The term **isotonic** means that there is an equal amount of **solute** both inside and outside the cell (therefore, the relative amounts of water on either side of the cell membrane are also equal). This is an equilibrium state; however, water does move equally in either direction. **Hypotonic** means less solute (more water) relative to another solution. **Hypertonic** means more solute (less water) relative to another solution. If a cell is placed in a hypotonic solution (i.e., one with more water) water will move into the cell. The cell could swell and burst. If a cell is placed in a hypertonic solution (i.e., one with less water) water will move out of the cell and the cell will shrink.

To demonstrate osmosis:

1. Obtain a 12 cm piece of dialysis tubing which has been soaked in water (This tubing is a synthetic equivalent to the cell membrane.),
2. Tie one end of the tubing with dental floss so that it is watertight,
3. Separate the other end of the tubing and fill the tube just over one-half full with a 50% sucrose solution and tie the end, leaving air space above the solution, so that it is watertight,
4. Blot the tube with paper toweling, weigh and record the weight in Table 10.1,
5. Place the tube in a 500 ml (or larger) beaker which contains approximately 8 cm of water,
6. At 15-minute intervals, remove, blot, and weigh the tubing and record the weight in Table 10.1.

TABLE 10.1 WEIGHT OF DIALYSIS TUBING BAG OVER TIME

MINUTES	WEIGHT IN GRAMS
0	
15	
30	

ACTIVITY 3

Plasmolysis and Turgor Pressure (Osmosis in Living Cells—Elodea and Potato)

Plasmolysis occurs when a cell, bounded by a cell wall, is placed in a hypertonic solution. The cell inside the cell wall shrinks due to water loss. **Plasmolysis** is the shrinkage of a cell due to water loss. The cell walls do not shrink but do lose the support of pressure formerly exerted by the cytoplasm. The pressure of the cell cytoplasm on the cell membrane and cell wall is called **turgor pressure.** Plasmolysis results from a loss of turgor pressure. Turgor pressure is directly related to the amount of water in the cell and is a part of the structural support of plant cells. Loss of turgor in higher plants is called **wilting.**

To observe plasmolysis and turgor pressure in living cells:

1. Take two large test tubes and label one distilled water (d H_2O) and the other salt water (NaCl H_2O). Fill each tube approximately one-half full of the designated water and place a strip of potato in each tube. Set aside for later observation.
2. Prepare a "wet mount" of an *Elodea* leaf,
3. Observe the cells in the leaf with 4×, 10×, and 40× objectives and note the appearance of the cytoplasm and the distribution of the chloroplasts and describe in Table 10.2,
4. Add two or three drops of a 20% NaCl solution to the leaf (lift the cover slip) and observe with the 10× and 40× objectives. Record the changes in the cytoplasm of the cells and the distribution of chloroplasts in Table 10.2.
5. Blot excess water from the slide with paper toweling and place several drops of distilled water on the leaf. Observe the cytoplasm of the cells, noting any changes in the distribution of chloroplasts. Record these observations in Table 10.2.
6. Observe the strips of potato that have been soaking in distilled water and in salt water by holding the potato strips in your hands. Record your observations in Table 10.3.

TABLE 10.2 OBSERVATION OF *ELODEA* LEAF

TREATMENT	OBSERVATIONS
distilled water	
5% salt water	
distilled water	

TABLE 10.3 OBSERVATION OF POTATO STRIPS

TREATMENT	OBSERVATIONS
5% salt water	
distilled water	

NAME ________________________________

EXERCISE 10

QUESTIONS

1. Define the following:

   diffusion

   osmosis

   active transport

   plasmolysis

   hydrophobic

   hydrophilic

   turgor pressure

2. Using only the results from activity 2, which way did the water move?

3. What membrane process operates in your lung tissue with respect to the movement of oxygen and carbon dioxide?

4. Why must aquatic organisms, such as amoebae, constantly use a contractile vacuole to transport water out of the cell?

5. What accounts for the spread of the purple potassium permanganate color throughout the tube in the first activity?

6. According to your results from activity 3, is plasmolysis reversible?

7. After 60 minutes in distilled water, your dialysis bag filled with sugar water fails to gain additional weight. Explain.

EXERCISE 11

MITOSIS/MEIOSIS

INTRODUCTION TO MITOSIS

According to the **cell theory** the basic unit of life is the cell. As with all living organisms cells have a life cycle. In this case, we call it the "**cell cycle**" (Figure 11.1). The majority of the life cycle of the cell is **interphase**. This is the stage in the cell cycle where the cell is not actively dividing, and we can break this stage down into three separate stages in which various processes occur. However, the remaining part of the cell cycle is **mitosis**. This is the process where the cell divides. In eukaryotic cells this process has four stages: **prophase, metaphase, anaphase,** and **telophase. Prokaryotic** cells divide by a different process called **binary fission**. The basic

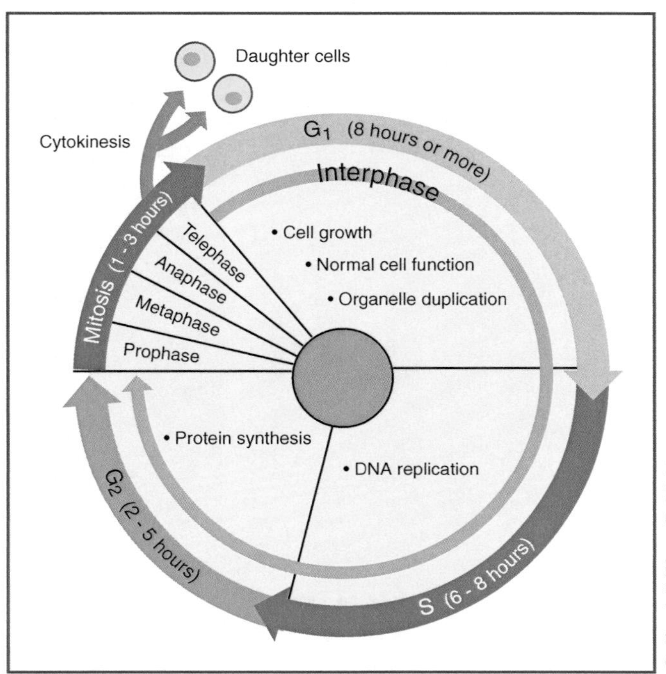

FIGURE 11.1

process is this: A **diploid** cell divides itself in such a way that it creates two genetically identical diploid cells. A diploid cell is a cell with a full set of chromosomes and is abbreviated **2n**. The diploid number for humans is 2n = 46, which means that within our cells we have 46 chromosomes (except the sex cells).

MITOSIS IN ANIMAL CELLS

Mitosis is the process where a diploid cell divides, creating two genetically identical diploid cells. There are four stages during mitosis: prophase, metaphase, anaphase, and telophase (Figure 11.2), and they occur in that order. Just remember the phrase "Pay me anytime" (PMAT), and you will never forget the correct order.

Prophase

Normally **chromosomes** are not visible as they are long thin strands of DNA; however, during this stage of mitosis they coil up and become thicker so they are visible (Figure 11.3). Also, the **nuclear membrane** and **nucleolus** disappear so they are no

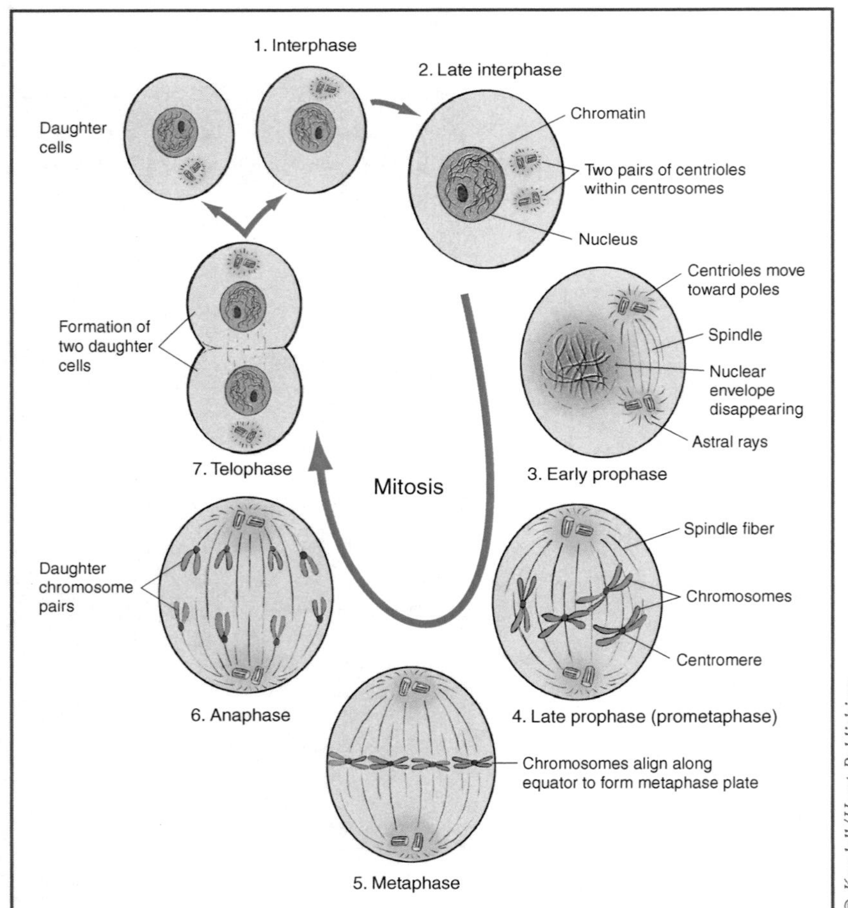

FIGURE 11.2

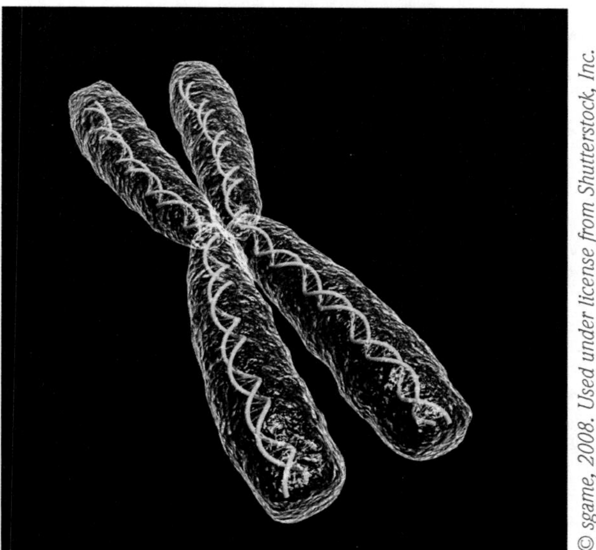

FIGURE 11.3

longer visible. The **centrioles** migrate toward opposite ends of the cell, and **spindle fibers** radiate between them. Figure 11.2 illustrates this. At this point it is important to understand how a chromosome is put together. Each chromosome consists of two **chromatids**, which are exact copies of each other (Figure 11.3). These two sister chromatids join at the **centromere**.

Metaphase

During this stage each chromosome is attached to two spindle fibers, one from each centriole. These spindle fibers are attached at the centromere. Also, the chromosomes are all lined up along the equator forming the **metaphase plate** (Figure 11.2).

Anaphase

During this stage the chromosomes split at the centromere and the chromatids pull apart. This gives rise to two **daughter chromosomes**, which are pulled toward opposite ends of the cell. Each end of the cell receives the diploid number. Also, because sister chromatids are exact copies of each other the two ends of the cell receive the exact same genetic information.

Telophase

This is essentially the reverse of prophase. During this stage the nuclear membrane and nucleolus reappear, and the chromosomes become less visible as they uncoil. Also, the cell divides in half. This process is called **cytokinesis** and occurs during telophase. In animal cells a **cleavage furrow** is produced as the plasma membrane of the cell pulls inward toward the center. Two new diploid cells form once the sides of the plasma membrane join in the center.

EXAMINATION OF THE WHITEFISH BLASTULA

Examine a prepared slide of the whitefish blastula and locate all stages of mitosis. Draw what you see in the space provided and label all structures that you can identify.

Prophase

Metaphase

Anaphase

Telophase

MITOSIS IN PLANT CELLS

Mitosis in plant cells occurs in essentially the same way as it does in animal cells; however, there are a few differences. First of all there are no centrioles. Instead the spindle fibers radiate from a **centrosome**. The centrosome organizes **microtubules** to form the spindle. Cytokinesis is also different in plant cells. Instead of the plasma membrane pulling in toward the center and forming a cleavage furrow a **cell plate** forms. This begins in the center of the cell and grows out toward the edges of the cell, forming a new **cell wall** that splits the original cell in half.

EXAMINATION OF THE ONION ROOT TIP

Examine a prepared slide of the onion root tip and locate all stages of mitosis. Draw what you see in the space provided and label all structures that you can identify.

Prophase

Metaphase

Anaphase

Telophase

INTRODUCTION TO MEIOSIS

One question that might arise is how do we maintain continuity of chromosome number between generations? If a sperm cell with 46 chromosomes fertilizes an egg with the same number then our cells should have 92, and that number should increase the next generation. The answer to this problem is meiosis. Sperm and egg cells are not diploid; rather they are **haploid**. A haploid cell is one that has half of the number of chromosomes; its abbreviation is **1n**. The haploid number for humans is 1n = 23. Now, when two haploid cells (sperm and egg) join they form a diploid

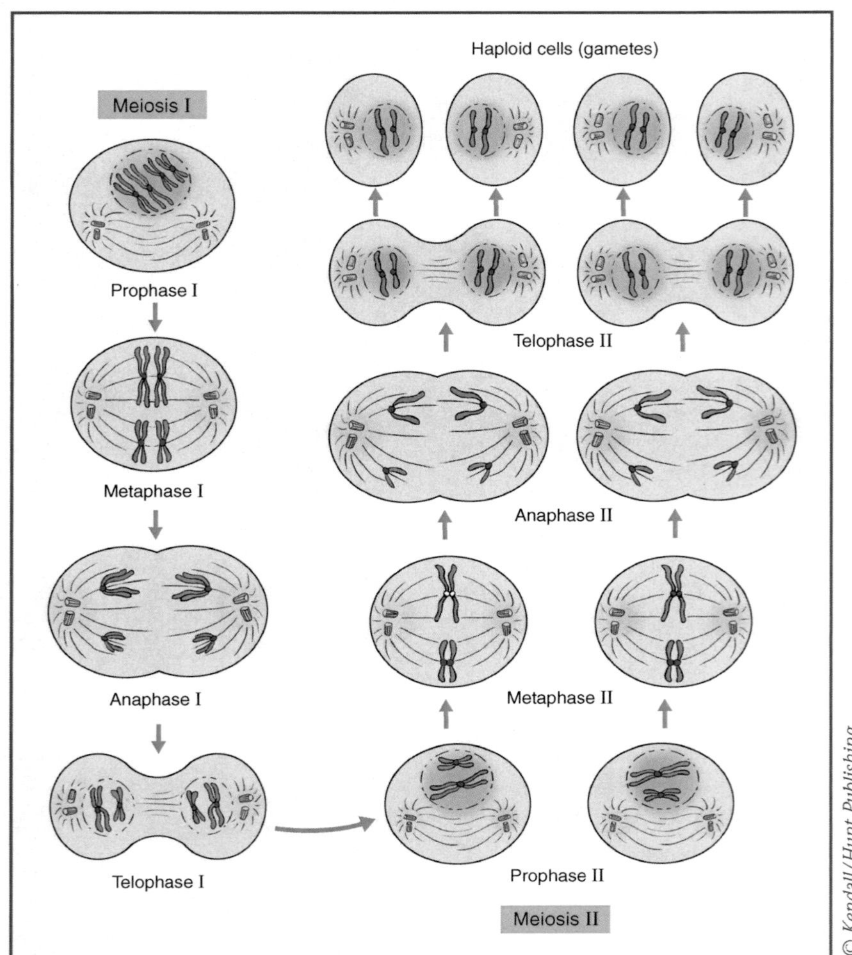

FIGURE 11.4

cell. In this way the number of chromosomes remains constant from one generation to the next, which is necessary in sexually reproducing species. Meiosis consists of two divisions (Meiosis I and II) with four stages each. Figure 11.4 shows this process.

Meiosis I

Prophase I

During this stage the nuclear membrane and nucleolus disappear, and the chromosomes coil and thicken so they become visible. Also, the centrioles migrate toward opposite ends of the cell, and spindle fibers radiate between them. This process is essentially the same as prophase of mitosis with the following exceptions. During this stage **homologous chromosomes** pair up, forming a chromosome tetrad (Figure 11.5). This is called **synapsis**. Homologous chromosomes have genes for the same trait. One comes from your mother and the other comes from your father. There are 23 pairs of homologous chromosomes in a diploid cell. When homologous chromosomes are in synapsis **crossing over** occurs. This process occurs when homologous chromosomes exchange pieces (Figure 11.5), which increases genetic

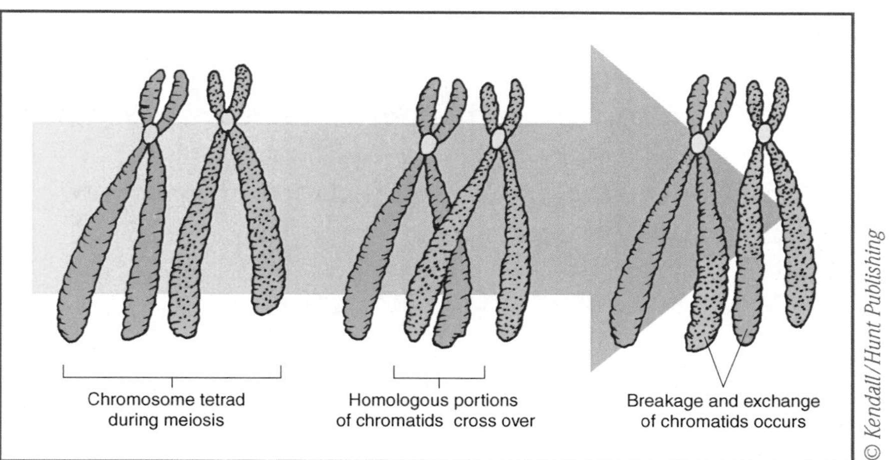

FIGURE 11.5

diversity. The X-shaped connection between homologous chromosomes during crossover is called a **chiasma**.

Metaphase I

During this stage homologous chromosomes line up along the equator of the cell with each pair attached to the spindle fibers. However, unlike mitosis, each chromosome is attached to only one spindle fiber (Figure 11.4).

Anaphase I

During this stage homologous pairs separate as each member of the pair is pulled toward opposite ends of the cell.

Telophase I

During this stage the cell divides in half. Each cell now contains half of the chromosomes of the original cell. The two cells are also genetically different because of crossing over and the separation of homologous chromosomes.

Meiosis II

Prophase

Chromosomes coil up and become visible. The nuclear membrane and nucleolus disappear. The centrioles migrate toward opposite ends of the cell, and spindle fibers radiate between them (Figure 11.4).

Metaphase

During this stage each chromosome is attached to two spindle fibers: one from each centriole. These spindle fibers attach at the centromere. Also, the chromosomes are all lined up along the equator, forming the metaphase plate (Figure 11.4).

Anaphase

During this stage the chromosomes split at the centromere and the chromatids and are pulled apart. This gives rise to two daughter chromosomes that are pulled toward opposite ends of the cell. Because of crossing over during prophase I the sister chromatids are no longer exact copies of each other so the two ends of the cell receive different genetic information.

Telophase

Cytokinesis occurs, creating two genetically different haploid cells from each haploid cell that began meiosis II.

COMPARISON OF MITOSIS AND MEIOSIS

It is clear from the description of meiosis that the second division (Meiosis II) is essentially mitosis. The only difference is that the cells are haploid instead of diploid, and they are not genetically identical. Figure 11.6 simultaneously compares mitosis and meiosis.

Take a cell (2n = 6) and draw all stages of mitosis and meiosis in the space provided. Be sure and label the chromosomes, centrioles, spindle fibers, centromere, chromatids, nuclear membrane, and nucleolus.

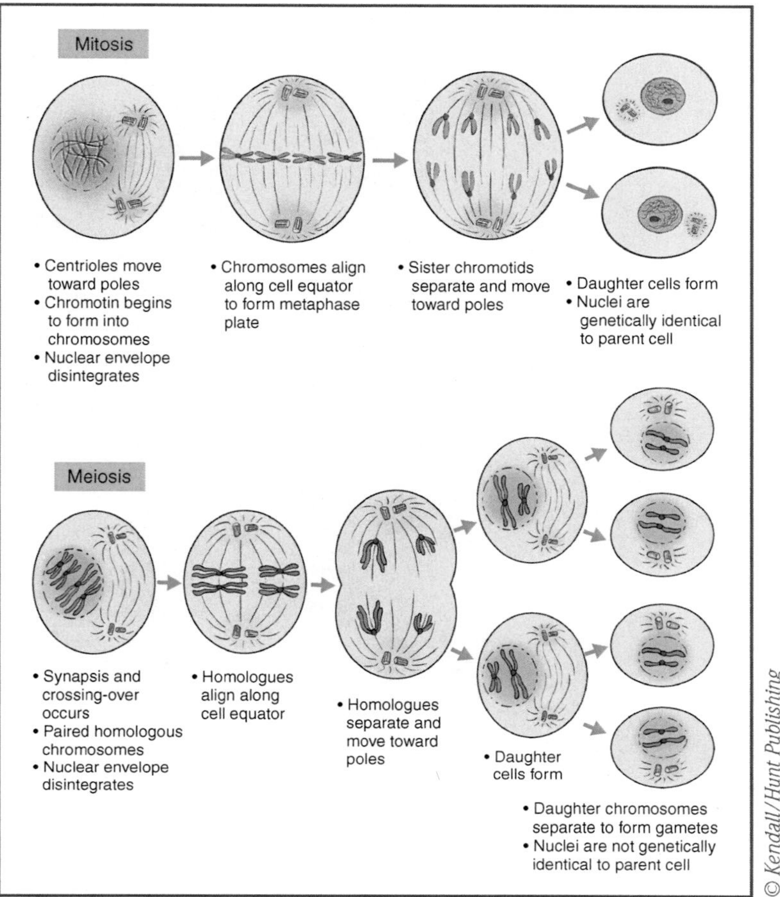
Mitosis
• Centrioles move toward poles
• Chromotin begins to form into chromosomes
• Nuclear envelope disintegrates
• Chromosomes align along cell equator to form metaphase plate
• Sister chromotids separate and move toward poles
• Daughter cells form
• Nuclei are genetically identical to parent cell
Meiosis
• Synapsis and crossing-over occurs
• Paired homologous chromosomes
• Nuclear envelope disintegrates
• Homologues align along cell equator
• Homologues separate and move toward poles
• Daughter cells form
• Daughter chromosomes separate to form gametes
• Nuclei are not genetically identical to parent cell

FIGURE 11.6

NAME ____

EXERCISE 11

QUESTIONS

1. Use the space below to draw and label all stages of mitosis (in order) for a diploid cell of 8.

2. Use the space below to draw and label all stages of meiosis (in order) for a diploid cell of 8.

EXERCISE 12

GENETICS

INTRODUCTION

Ever wondered why you have your father's eyes but your mother's nose? A field of biology called **genetics** answers questions of inheritance. The field of genetics is growing rapidly and will continue to do so; however, genetics owes a debt of gratitude to a nineteenth-century Austrian monk named **Gregor Mendel**. Mendel worked with garden peas in the monastery's garden. He crossed plants with certain characteristics such as pea color and shape to see what the offspring looked like. In today's lab you will learn how traits are passed from parent to offspring.

THE PUNNETT SQUARE

The first step in any genetics problem is to create a **Punnett square**. A Punnett square makes predictions concerning the offspring from a mating between two individuals. The first step is to determine the parents' **genotypes**. The genotype refers to the **alleles** that an organism has for a trait. Alleles are alternative forms of a gene. If an organism is **homozygous** that means that both alleles for a trait are the same (two alleles for black hair). If an organism is **heterozygous** that means that the two alleles are different (one allele for black hair and one for brown hair). The next step is to put the **gametes** (sperm and egg) in a Punnett square. Remember that the gametes will carry only one allele because they are **haploid**. Figure 12.1 shows this step and uses two heterozygous parents. The next step is to fill in the Punnett square. Figure 12.1 also shows this. The completed Punnett square shows all possible offspring from a mating from parents with certain genotypes.

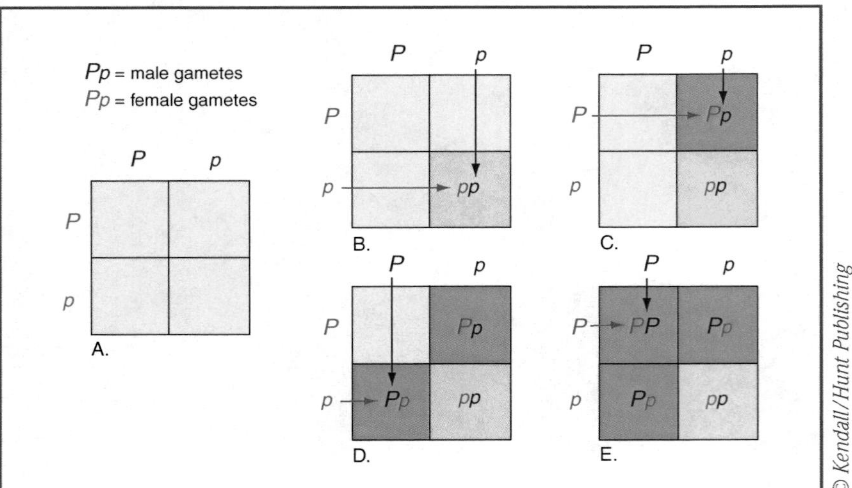

FIGURE 12.1

THE MONOHYBRID CROSS

In a **monohybrid cross** all you are interested in is the inheritance of a single characteristic, for instance hair color or eye color. Suppose you had a purple flower and a white flower. If all you are interested in knowing is the color of the flowers of the offspring from the white and purple flower parents that is a monohybrid cross. In any cross it is helpful to know which allele is **dominant** and which is **recessive.** A dominant allele will express itself in either the homozygous or heterozygous genotype. A recessive allele will express itself only in the homozygous condition. We typically write dominant alleles in capital letters and the recessive allele in lowercase letters. Examine Figure 12.2. This shows what a cross between a purple and a white flower would look like. Assume that purple is dominant to white. If the purple flower is homozygous (PP) and mates with a white flower (pp), then all offspring will have a heterozygous (Pp) genotype and the phenotype will be purple. **Phenotype** refers to the physical characteristics of an organism; in this case the phenotype is a purple flower. However, if both purple flowers are heterozygous (Pp) then there is a 75% chance of having a purple flower and a 25% chance of having a white flower. We can determine these ratios by looking at the Punnett square. If three out of the four possibilities are purple, then that is a 75% chance of purple flowers.

Examine Figure 12.3. This shows the monohybrid cross in more detail. The P generation refers to the original parents. The P generation produces the F1 generation, and mating individuals of the F1 generation produces the F2 generation. In this example (Figure 12.3) we have guinea pigs with black and brown fur mated to one another where black is dominant to brown.

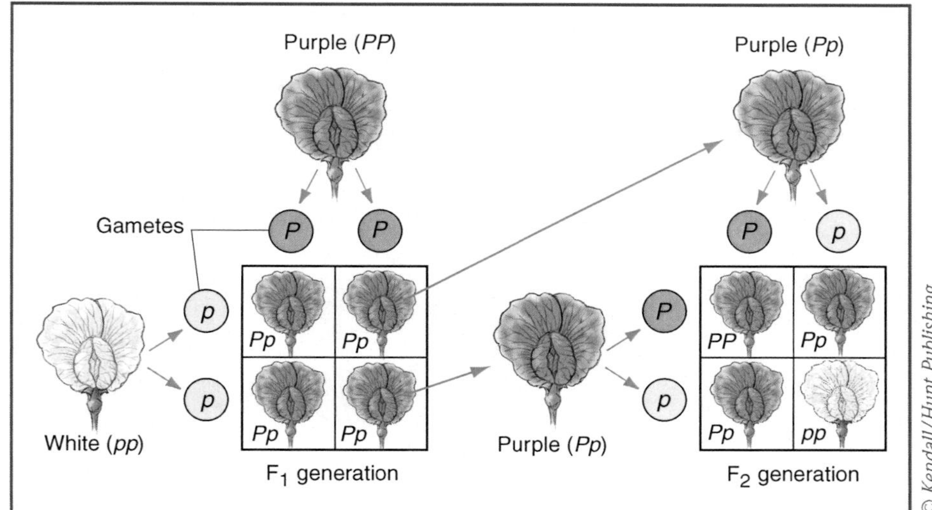

FIGURE 12.2

P generation

Black female (*BB*)
Brown male (*bb*)

Gametes produced by P generation

B
Dominant *B* masks recessive *b*
b

Female (*Bb*)
Male (*Bb*)

F1 generation (all *Bb*)

Gametes produced by F1 generation

1/2 *B*
1/2 *b*
1/2 *B*
1/2 *b*

Combinations of F1 gametes to produce F2 generation

F2 genotypic ratio = 1 *BB* : 2 *Bb* : 1 *bb*

	B	*b*
B	1/4 *BB*	1/4 *Bb*
b	1/4 *Bb*	1/4 *bb*

FIGURE 12.3

Practice Monohybrid Genetics Problems

Use the space provided to answer the following problems.

1. Red eyes are dominant to yellow eyes. Give the genotypes and phenotypes of the offspring you can expect from a cross between two parents with the following genotypes. Parent one is heterozygous for eye color, and parent two is homozygous for the recessive eye color.

2. If two heads (H) are dominant to one head (h), what would the F1 and F2 generation look like (what is their phenotype) if the P generation is HH × hh?

THE DIHYBRID CROSS

In a **dihybrid cross** you are interested in tracking the inheritance of two characteristics. Figure 12.4 shows a dihybrid cross. In this example we are interested in coat color and hair length. The trick to a dihybrid cross is to put the right gametes in the Punnett square. If an organism if heterozygous for two characteristics, it has the genotype "AaBb" and can make the following gametes: AB, Ab, aB, ab. Notice that there are no gametes like "Aa" or "Bb." This is because the "A" and "a" are on **homologous chromosomes** and separate during **anaphase I** of **Meiosis**. The same is true for "B" and "b." Once you put the right gametes into the Punnett square, it works the same as a monohybrid cross except there are more possibilities for the offspring because you are tracking two characteristics instead of one (Figure 12.4).

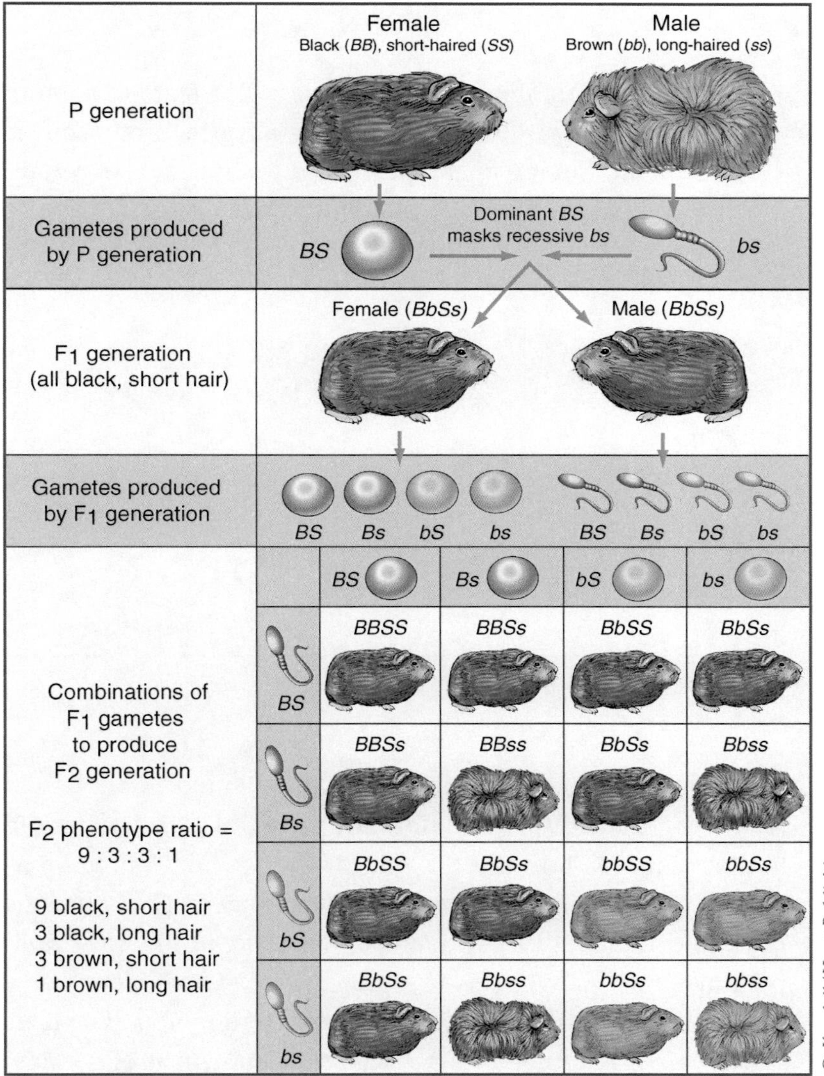

FIGURE 12.4

Practice Dihybrid Cross Problems

Use the space provided to answer the following problems.

1. What are all the possible gametes (sex cells) that parents with the following genotypes can make?

   BbFF

   bbGg

   AaBb

2. Red eyes are dominant to yellow eyes. Blue feet are dominant to green feet. Give the genotypes and phenotypes of the offspring from a cross between two parents with the following genotypes: Parent one is heterozygous for both characteristics, and parent two is homozygous for the recessive eye color and heterozygous for foot color.

THE TESTCROSS

Suppose you found a purple flower and wanted to know what the genotype was. Assuming that purple is dominant to white, the one you found could either be homozygous (PP) or heterozygous (Pp), but how can you know for sure? You will have to do a testcross. In a **testcross** you mate your organism with an organism that has a known genotype for that characteristic. In this case you would mate your purple flower with a white flower. You already know what the genotype of the white flower is because the only way to have a white phenotype is to be "pp." Figure 12.5 shows this. You can determine what the genotype is of the flower you found by looking for any white offspring. If there are many offspring produced and all are purple,

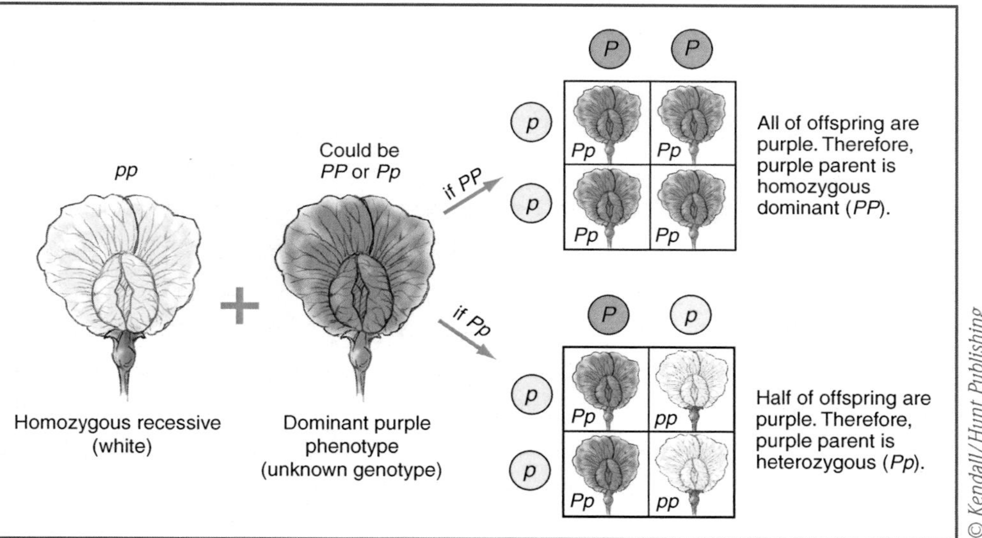

FIGURE 12.5

then the genotype of the flower you found is "PP," but if any white flowers show up then the genotype must be "Pp" (Figure 12.5).

Practice Testcross Problems

Webbed feet are dominant to non-webbed feet. You come across an animal that has webbed feet and decide to determine what this organism's genotype is.

1. What are the possible genotypes of the organism that you found?

2. What is the genotype and phenotype of the organism you would use to cross with the animal you found?

3. What is the genotype of the organism you found given the following results of the cross:

   a. 2 offspring with webbed feet

   b. 1 offspring with non-webbed

SEX-LINKED TRAITS

Within the cells of your body (except sex cells) you have 23 pairs of homologous chromosomes. Twenty two of these pairs are **autosomes** and one pair of **sex chromosomes**. Sex chromosomes determine what sex you are. Females have two "X" chromosomes, and males have one "X" and one "Y" chromosome. Traits carried on sex chromosomes are sex-linked. That is, they link to a sex chromosome. Most sex-linked traits are located on the "X" chromosome as the "Y" chromosome carries very little except what makes a person a male. For females this is not a problem because normal rules apply; however, this is not true for males. Because males only have one copy of the "X" chromosome they will express whatever they receive. For instance, for a female to be colorblind (a sex-linked trait) she must receive two copies of the recessive allele, but males only need to receive one copy since there is nothing on the "Y" chromosome to mask the effects of the recessive allele. In other words all he has to have is one recessive allele to be colorblind instead of two. Sex-linked traits are typically written on the "X" chromosome, and the genotypes are shown below for colorblindness.

Normal vision female	X^AX^A	or	X^AX^a
Colorblind female	X^aX^a		
Normal vision male	X^AY		
Colorblind male	X^aY		

Practice Sex-Linked Problems

If a colorblind female marries a normal vision male and has children, what are the chances of the following offspring?

1. Normal vision son

2. Normal vision daughter

3. Colorblind son

4. Colorblind daughter

MULTIPLE ALLELES

So far all we have considered are situations where there are only two alleles for a trait. However, there may be several alleles for any one characteristic. An example of this is the blood groups. There are a total of three alleles for human blood types: A, B, and O. Realize that while there are three alleles you can only possess two at any one time. We write the genotypes for the different blood types as follows:

Type A blood:	I^AI^A	or	I^Ai
Type B blood:	I^BI^B	or	I^Bi
Type AB blood:	I^AI^B		
Type O blood:	ii		

Notice that both the A and B alleles are dominant to the allele for O blood. Also notice that someone who possesses both the A and B alleles has AB type blood. This is a situation where the two alleles are co-dominant. The phenotype simultaneously expresses both co-dominant alleles.

What are all the possible genotypes of the parents if their child has "O" type blood?

NAME ____________________

EXERCISE 12

QUESTIONS

1. **Animal 1:** Brown fur with no tail
   Animal 2: Black fur with no tail

   Brown is dominant to black, and no tail is dominant to having a tail.
   Animal 1 is heterozygous for both characteristics.
   Animal 2 is heterozygous for no tail.

   Mate these two individuals and answer the following questions.

   a. What are all the possible genotypes from a cross between these two individuals?

   b. What are all the possible phenotypes from a cross between these two individuals?

2. A dominant allele genetically controls the ability of humans to taste the bitter chemical phenylthiocarbamide (PTC). People with at least one copy of the dominant allele of this gene can taste PTC; those who are homozygous recessive for the non-tasting allele cannot taste it.

   a. Could two parents able to taste PTC have a child who cannot taste PTC? Explain.

   b. If two parents, both of which had a taster and non-taster parent, can taste PTC and are expecting their first child, what are the chances that their child will be able to taste PTC? What are the chances that their child will be a non-taster?

EXERCISE 13

CONTROL WITHIN CELLS

INTRODUCTION

In the field of Biology, models are often used to visually demonstrate concepts or principles. A model can help us integrate known facts into a complete conceptual picture, much like assembling the pieces of a picture puzzle. Such was the technique used by Watson and Crick to demonstrate the probable configuration of **deoxyribonucleic** acid (DNA).

In this exercise, students will construct part of a DNA molecule from parts provided in the kits. Before beginning, it must be understood that the nucleic acid DNA consists of two helical strands, each made up of a series of nucleotides. Each of the nucleotides contains three molecules: a **nitrogenous base**; the 5-carbon sugar, **deoxyribose**; and a **phosphate** molecule. The sugars of the individual nucleotides are linked through the phosphate units to form the backbone of the strand. The two strands are held together by hydrogen bonds that bond the nitrogenous bases of one strand to those of the other strand. Two kinds of nitrogenous bases are in DNA: purines and pyrimidines. Purines are always united by hydrogen bonds to pyrimidines, forming a cross bridge between the two nucleotide strands in the DNA molecule. The purine **adenine** normally forms hydrogen bonds only with the pyrimidine **thymine**; the purine **guanine** normally forms hydrogen bonds only with the pyrimidine **cytosine**.

The production of two identical DNA molecules from one template (model) molecule is accomplished by what is referred to as **semi-conservative replication**. The process results in two new DNA molecules each consisting of one old strand and one new strand. This formation may be studied at three stages. First, the nucleotide strands separate as the hydrogen bonds connecting them are ruptured. Then new nucleotides form hydrogen bonds with complementary nucleotides in both of the original strands. Adenine always bonds with thymine; cytosine always bonds with guanine. Finally the new nucleotides are polymerized (bonded together).

To understand protein synthesis, the student should be familiar with the structure and function of another nucleic acid: **Ribonucleic acid** (RNA). RNA, which is formed as a complementary copy of one of the DNA strands by a process called **transcription**, differs only in minor ways from DNA. The following table summarizes these differences:

RNA	DNA
Sugar (5 carbon) ribose	Sugar (5 carbon) deoxyribose
Uracil (one pyramidine base)	Thymine (one pyrimidine base)
Single strand	Double strand
Found in the nucleus and in the cytoplasm	Found in the nucleus as part of the chromosomes, in mitochondira and chloroplasts, perhaps in centrioles

There are three specific kinds of RNA: **messenger RNA, ribosomal RNA** and **transfer RNA**. Present evidence indicates that only one DNA strand is involved in RNA manufacture. The RNA molecules leave the nucleus where they are formed and move into the cytoplasm. Ribosomal RNA (rRNA) molecules become associated with a ribosome in the cytoplasm. Each ribosome may form a complex with other ribosomes to produce polysomes. The messenger RNA (mRNA) associates with the ribosomal RNA and the exposed nitrogenous bases on the mRNA will, by their sequence, determine which transfer RNA molecules will pair with the mRNA. Beginning at one end of the mRNA, successive groups of three bases (triplets) represent a "code word" or codon for a specific amino acid. The transfer RNA units are, like other kinds of RNA, a single strand, but they coil back on themselves and hydrogen bonds form between appropriate base pairs, except at one end where three nitrogenous bases are not paired with others of the same strand. It is these three bases, called an **anti-codon**, which form hydrogen bonds with the complementary bases on the mRNA molecule.

Each transfer RNA unit, due to its specific structure, has an affinity for a specific amino acid, when activated by the appropriate enzymes. Once the transfer RNA unit has united with the appropriate amino acid, it will be "attracted" to the specific bases along the messenger RNA molecule. The exposed transfer RNA bases will align with complementary bases on mRNA and hydrogen bonds will form between these base pairs. Once the series of nitrogenous bases in the messenger RNA molecule have formed, hydrogen bonds with nitrogenous bases in the appropriate transfer RNA molecule (each of which is associated with a specific amino acid), peptide bonds form between the amino acids thus brought into close association, forming a specific polypeptide chain. Once the amino acids are joined by the peptide bonds, they lose their affinity for their transfer RNA molecules. The transfer RNA molecules in turn separate from the bases of the messenger RNA. As long as the specific enzymes are present the entire sequence of events will be repeated until a sufficient number of molecules of the specific protein have been formed. This process whereby the RNA code is read to form a protein molecules is called **translation.**

Mutations constitute the principle raw material with which nature works to bring about evolution. Mutations are sudden, heritable changes in the structure of the genetic material. Most mutations are detrimental and recessive (must be present in double dose to be expressed). **Point mutations** occur as one type of nitrogenous base in DNA is replaced by another base. This results in a new codon which ultimately causes an incorrect amino acid to be substituted in the polypeptide chain. In order to observe how changes (mutations) in DNA can affect both the genetic inheritance in new cells and activities within cells, we will use the models to demonstrate the effect of mutations on DNA structure.

OBJECTIVES

After completing this exercise the student should be able to:

A. Describe the structure of the nucleic acids deoxyribonucleic acid (DNA) and ribonucleic acid (RNA).

B. Explain the mechanism by which DNA replicates.

C. Explain the mechanism of transcription and the process by which proteins are synthesized and the genetic information in the DNA structure is translated into protein structure.

D. Define point mutations and explain how they affect the structure and function of proteins.

E. Understand procedures that cause disruption of a yeast cell and release of contents.

F. Collect DNA from the released cell contents.

PROCEDURE

Construct the length of DNA diagramed below using your DNA kit and the following code:

A = deoxyribose
B = ribose
C = phosphate
D = uracil
E = thymine
F = cytosine
G = guanine
J = adenine
H = H + (from amino group)
OH = OH^- (from carboxyl group)

in DNA
J forms hydrogen bonds with E
F forms hydrogen bonds with G

in RNA
J forms hydrogen bonds with D
F forms hydrogen bonds with G

Always use the parts with the code letters up (towards you). Your instructor will provide any assistance you may need.

A. Using the left strand as a guide, construct a DNA molecule.

Left DNA Strand

S — A = P

S — C = P

S — T = P

S — G = P

S — A = P

S — C =

B. Replicate the DNA molecule as it is done in the cell.

C. Again, using the left strand as a template, construct a mRNA molecule (codons).

D. Following the directions on the mRNA codon construct tRNA molecules (anti-codons).

E. Attach an amino acid to the tRNA molecules, split out H_2O and form a peptide bond joining the amino acids.

F. Using the original DNA left strand, form a mutated mRNA strand by changing one of the bases.

YEAST DNA EXTRACTION

A. Procedures

1. Mix 1 package of dry yeast with 40 ml of 50°C hot tap water to dissolve the yeast in a beaker. Keep mixture covered and warm for about 20 minutes.
2. Add 40 ml detergent/salt solution.
3. Place mixture in a blender and blend 30 sec. to 1 minute on high.
4. Pour mixture back into the beaker, add 15 ml of meat tenderizer solution, and stir to mix.
5. Place 6 ml of mixture into a test tube.
6. Pour 6 ml of ice cold ethanol carefully down the side of the tube to form a layer.
7. Let the mixture sit undisturbed 2–3 minutes until bubbling stops.
8. A precipitate will form in the alcohol. Swirl a glass stirring rod at the interface of the two layers. The precipitate is DNA.

B. Results

1. Describe the precipitated DNA. ______________________________

   ____

   ____

2. Is this pure DNA? ______________________________

C. Questions

1. What causes the lysis of the cell wall and nuclear membrane of the yeast cell? ______________________________

   ____

2. What substance causes the precipitation of DNA? ______________________________
3. Can other organisms/cells be substituted for DNA extraction? ______________
4. Would the extraction procedure need to be changed? ______________

MATERIALS

DNA kits. Available from:

K. D. Biographies
1050 Flake Drive
Palatine, Illinois 60067

Dry yeast, Adolph's natural meat tenderizer (to make solution; 5 gm tenderizer in 95 ml distilled water), beakers distilled water, non-iodized salt, Palmolive detergent (to make detergent/salt solution; 20 ml detergent, 20 ml of non-iodized salt in 180 ml of distilled water), glass stirring rod, 10 ml and 100 ml graduated cylinders, blender, 15 ml test tubes, ice cold 95% ethanol, test tube rack.

NAME ____________________

EXERCISE 13

QUESTIONS

1. What parts of the DNA molecule repeat down the length of the strand?

2. What are the three component parts of each of the nucleotides?

3. Explain how the nucleotides are connected to each other in a single strand.

4. How are complementary strands connected to each other?

5. In DNA replication, hydrogen bonds form between certain base pairs. Name these pairs.

6. Which part of your DNA model is similar to coded information?

7. Would the RNA molecules be the same if copies were made from both nucleotide strands of DNA? Explain.

8. What happens to this RNA after being assembled in the nucleus?

9. How many amino acids are coded for by your messenger RNA model?

10. If a messenger RNA sequence reads adenine-uracil-cytosine-adenine-uracil-guanine, what is the nucleotide sequence of the two tRNA molecules that are complementary?

11. How does the sequence of nucleotides in transfer RNA compare with the sequence of nucleotides in the DNA from which you assembled the messenger RNA model?

12. DNA is important in the cell because it:

a. is a storehouse of genetic information which must be "handed down" to future generations of cells. How is its double stranded structure suited for the precise replication of this information prior to cell division?

b. controls the chemical activities of cells through synthesis of proteins (enzymes, antibodies. pigments, etc.). How is the linear sequence of nitrogenous bases along the DNA strand uniquely suited to encode information determining the linear sequence of amino acids in a protein (enzyme)?

13. How does changing the linear sequence of one of the DNA strands affect the newly synthesized DNA?

14. What affect will this have on new cells "descended" from this DNA?

15. What will be the effect of the altered mRNA on the sequence of amino acids in the protein synthesized from it?

16. What do you suppose would be the effect on the function of proteins made from mutated DNA molecules in which random changes are made in the amino acid sequence?

EXERCISE 14

PROTISTA

INTRODUCTION

The kingdom **Protista** is unique among kingdoms. While members of other kingdoms are grouped because of similar characteristics, the members of this kingdom are grouped because they belong nowhere else. Because of this you may be tempted to consider it a kingdom of rejects, but kingdom Protista is a highly diverse group with important ecological roles within the ecosystems in which it occurs. They are all **eukaryotic** but vary greatly in size from microscopic to the giant kelp forests. They can be beneficial (such as producing oxygen via photosynthesis) or harmful (such as malaria). The classification of this group is in a state of flux. Many scientists will disagree on the placement of members within this kingdom, and some will even put some members of this kingdom into other kingdoms. We will, however, rely on a more traditional view of this kingdom for simplicity. Today you will explore various representatives of this important group.

THE ALGAE

Algae are a diverse group within kingdom Protista. They range in size from a single cell to hundreds of feet long. You can find them in both aquatic and terrestrial habitats. Algae can also exist in a **symbiotic** relationship with **fungi** and form **lichens**.

Diatoms

These are unicellular algae that you can find in both freshwater and marine environments. In the marine ecosystem they form a large portion of the base of the **food chain**. They are easily recognized as they are symmetrical in shape and have shells made of silica. Figure 14.1 shows representatives of this group. Obtain a prepared

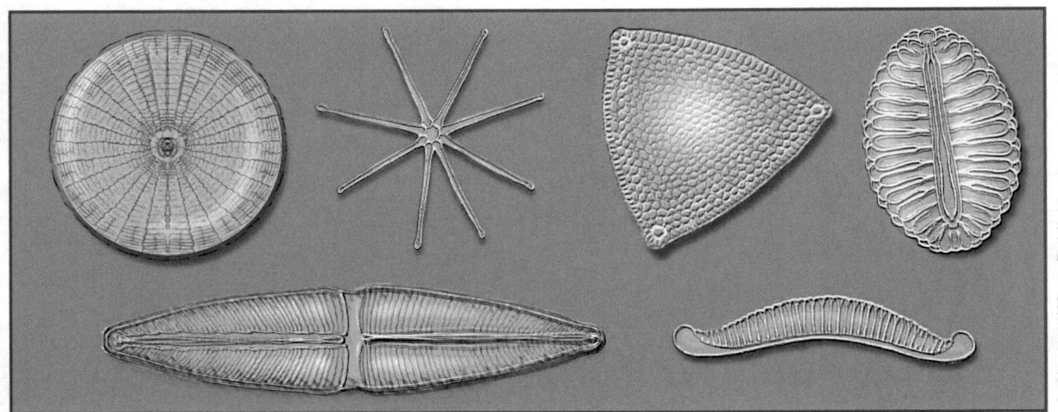

FIGURE 14.1

slide of diatoms and examine this with the aid of a microscope. Use the space below to draw what you see.

Dinoflagellates

These are unicellular and photosynthetic algae. Dinoflagellates are responsible for **red tides,** which occur when their populations greatly increase. They form symbiotic relationships with coral reefs and are **luminescent**. This causes a twinkling effect visible in disturbed water, such as in a ship's wake or during wave action. Dinoflagellates have two flagella (hence their name) and are easily recognizable because of their unique shapes (Figure 14.2). Obtain a prepared slide of dinoflagellates and examine the slide with the aid of a microscope. Use the space below to draw what you see.

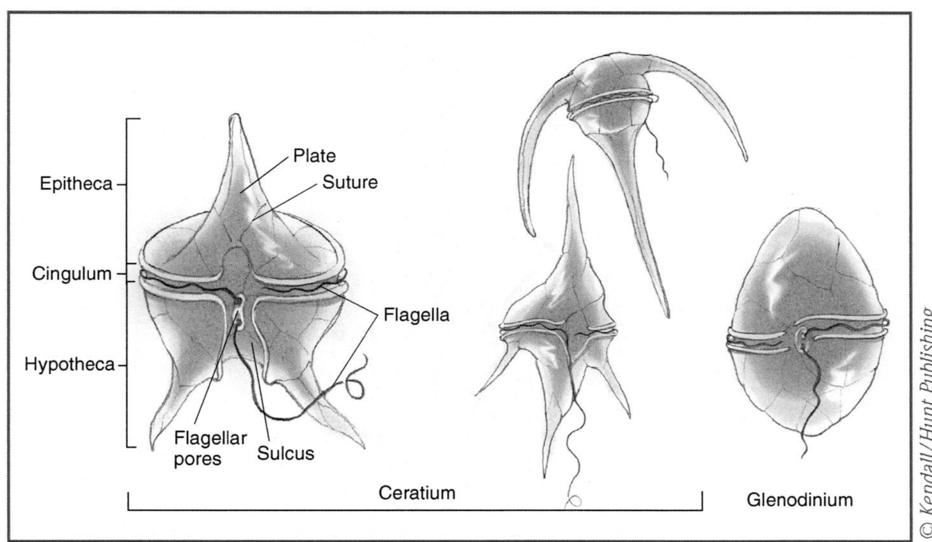

FIGURE 14.2

Green Algae

Green algae are closely related to the first plants since they share many of the same characteristics. Green algae may be unicellular or multicellular, and some green algae have symbiotic relationships with **fungi** to form **lichens**. Today we will examine two representative green algae: *Spirogyra* and *Volvox.*

Spirogyra consists of a chain of cells arranged end to end that contains strands of ribbon-like **chloroplasts** (Figure 14.3) that take on a spiral shape. Obtain a prepared slide of *Spirogyra* and examine it using the microscope. Use the space below to draw what you see.

Volvox is a multicelluar colony of flagellated cells with the "**daughter**" colonies developing inside the "**parent**" colony (Figure 14.4). Obtain a prepared slide of *Volvox* and examine it using the microscope. Use the space below to draw what you see.

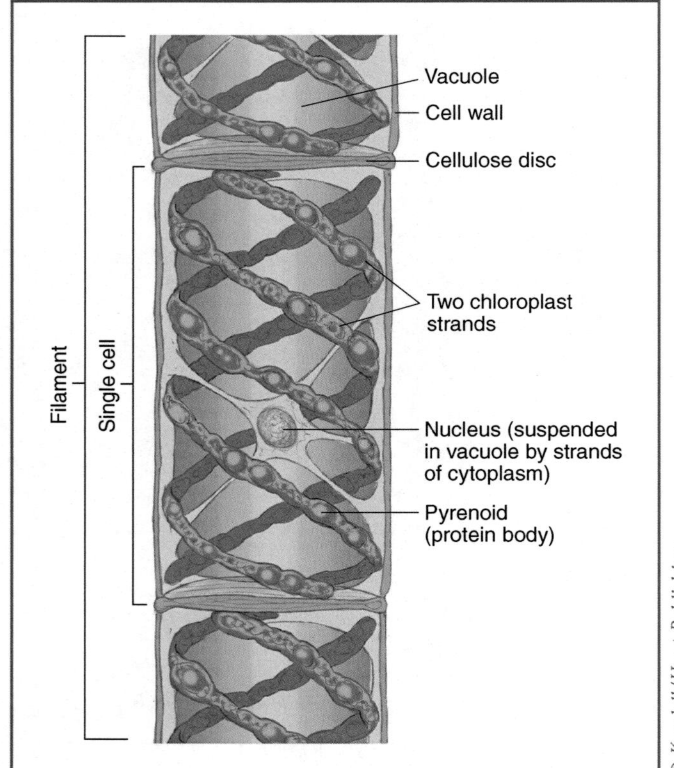

FIGURE 14.3

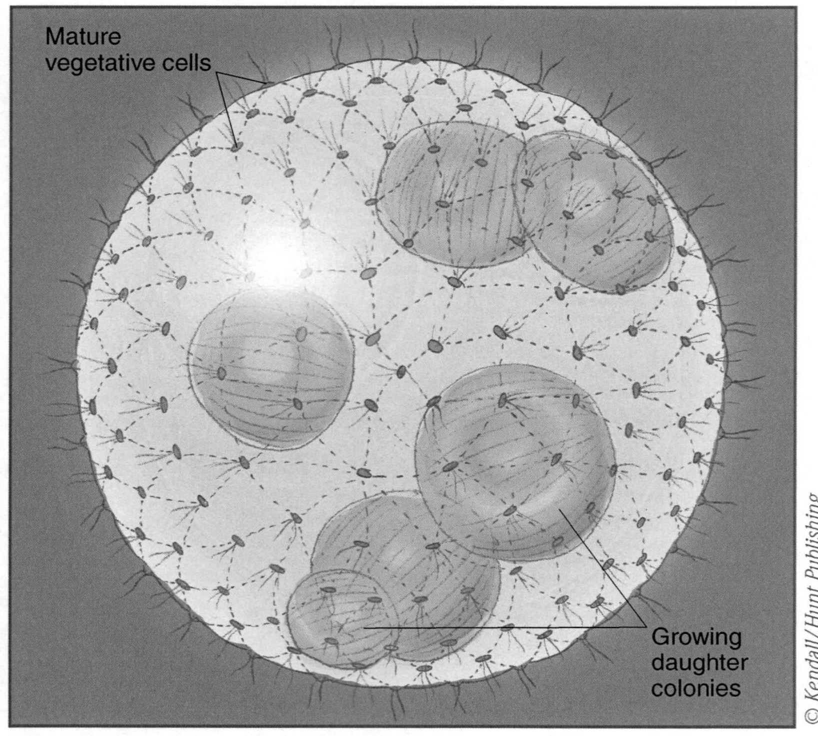

FIGURE 14.4

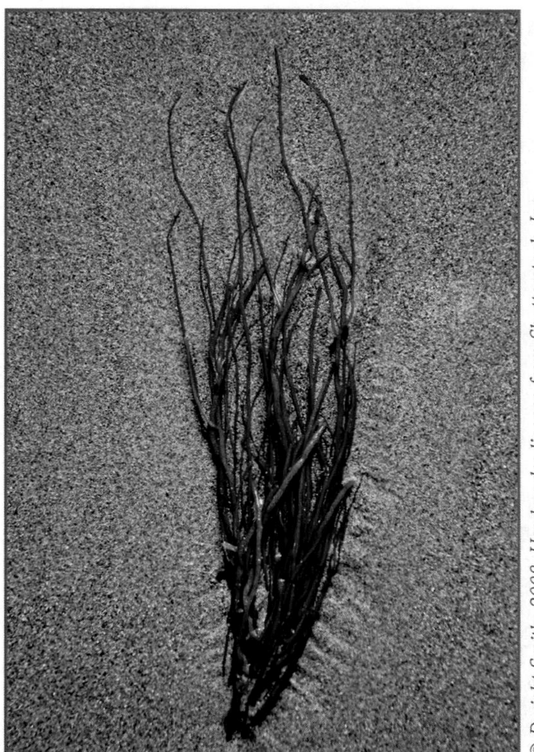

FIGURE 14.5

Red Algae

Red algae are primarily multicellular and can appear filamentous or leafy. They resemble brown algae; however, they are generally smaller and appear more delicate that brown algae. A representative red alga appears in Figure 14.5. Examine dried herbarium specimens of various red algae provided by your instructor.

Brown Algae

We find brown algae in cold or temperate marine habitats where they can grow to enormous size. The **Sargasso Sea** is famous for large floating mats of brown algae. The most famous examples of brown algae are likely the large kelp forests shown in Figure 14.6. Examine dried herbarium specimens of various brown algae provided by your instructor.

THE PROTOZOANS

The protozoans are unicellular eukaryotic protists. While they are unicellular they can be very complex. Some may even have more than one nucleus and reproduce sexually, although asexual reproduction is the most common method.

FIGURE 14.6

Euglenoids

Euglena is the best known member of this group. Many contain chloroplasts, but some do not. They also contain an **eyespot**, which is a photoreceptor, and two flagella, one of which is much longer that the other. Figure 14.7 shows a model *Euglena.* Obtain a prepared slide of *Euglena* and examine it using the microscope. Use the space below to draw what you see.

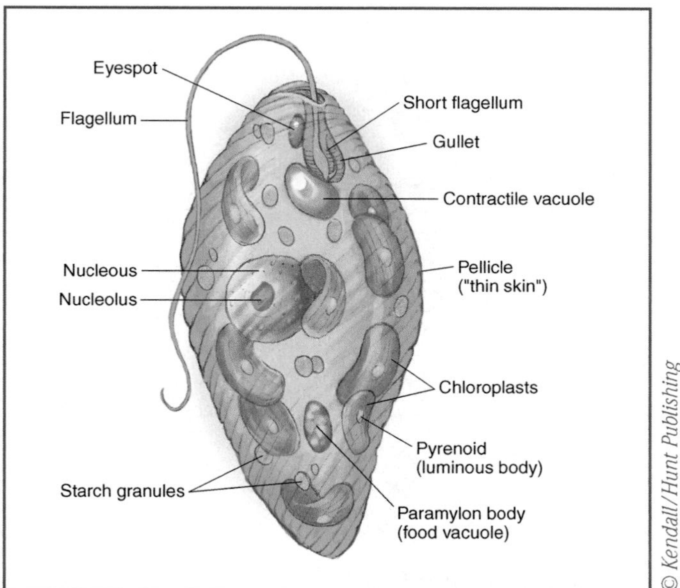

FIGURE 14.7

Ciliates

Paramecium best represent the ciliates (Figure 14.8). **Cilia** cover this organism, allowing the paramecium to move through its environment. As well as the typical eukaryotic organelles, *Paramecium* contains two nuclei (**macronucleus** and a **micronucleus**) and an **oral groove,** used during sexual reproduction called **conjugation.** Obtain a prepared slide of *Paramecium* and examine it using the microscope. Use the space below to draw what you see.

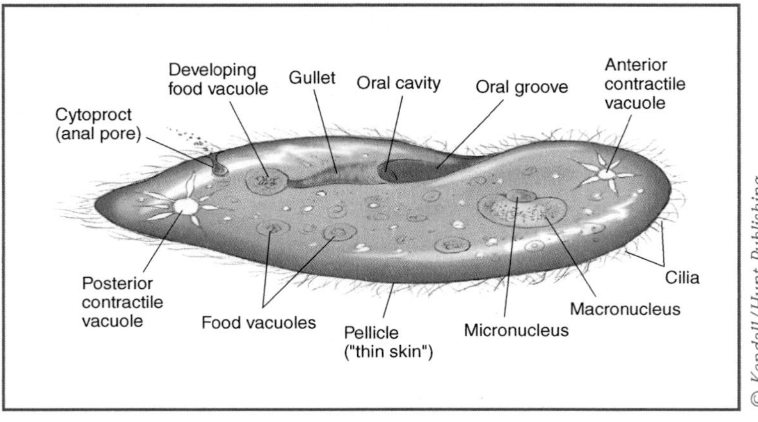

FIGURE 14.8

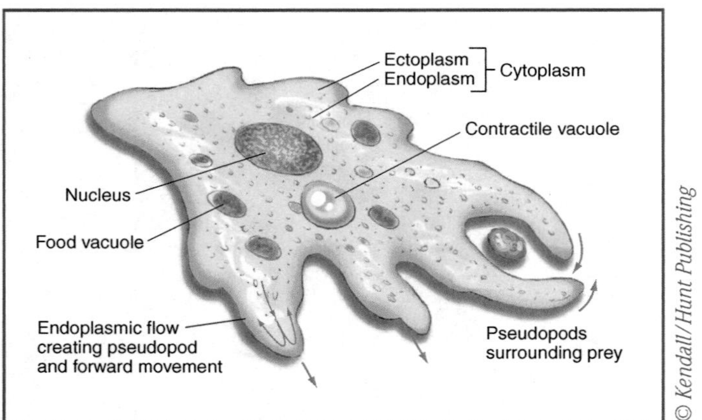

FIGURE 14.9

Amoeboids

These organisms, represented by the genus *Amoeba* (Figure 14.9), have an amorphous shape and move through the use of **pseudopodia**. Pseudopods form when their **cytoplasm** flows in a particular direction, causing their cell membrane to bend in that direction. Obtain a prepared slide of *Amoeba* and examine it using the microscope. Use the space below to draw what you see.

Zooflagellates

This is a group that has serious consequences for humans. Members of this group cause **African sleeping sickness**, **Chagas disease**, and **diarrhea**. Obtain a prepared blood slide with *Trypanosoma cruzi.* This is the zooflagellate that causes Chagas disease. Examine this under the microscope. Most of the structures you will see are red blood cells; however, you should also notice that there are some ribbon-like cells scattered among the red blood cells. This is *T. cruzi.* They contain an undulating membrane along their side that propels them through their environment and gives them that undulating shape.

Sporozoans

This is a group that gets its name because spores form. It is also the group that contains the biggest killer of humans since *Homo sapiens* first evolved: **malaria**.

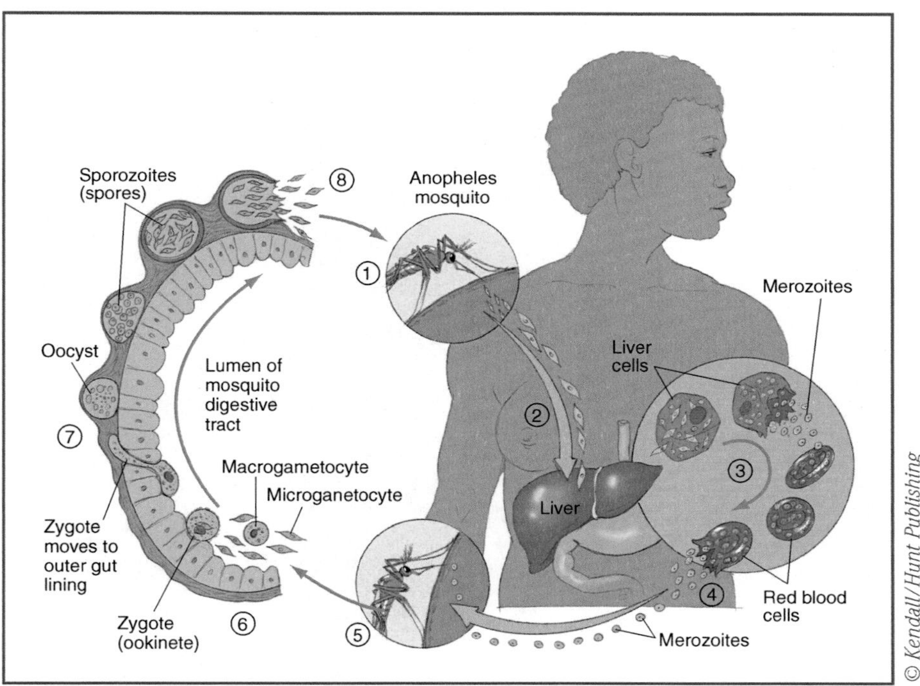

FIGURE 14.10

Figure 14.10 shows the life cycle of malaria. The symptoms of malaria include fever, vomiting, and anemia, which may result in death.

WATER AND SLIME MOLDS

Water molds are typically aquatic, and slime molds are found in terrestrial habitats. While molds are typically considered fungi, these two are protists. In the environments where water and slime molds appear, they feed on dead organic material and thus serve as **decomposers**.

EXAMINATION OF POND WATER

Make a wet mount using pond water provided by your instructor and locate as many members of the kingdom Protista as you can. Draw what you see.

NAME ____________________

EXERCISE 14

QUESTIONS

1. What Protistan is responsible for red tides?

2. Describe what *Spirogyra* looks like.

3. Which Protistan contains an eyespot?

4. What is unique about Kingdom Protista with regards to other kingdoms?

5. Draw a *Euglena.*

6. Draw a *Paramecium.*

EXERCISE 15

FUNGI

INTRODUCTION

Fungi are a diverse group of organisms that share similar characteristics. One characteristic they have is their mode of nutrition: They are strictly **heterotrophic**. That means that, unlike plants, they cannot make their own food but must obtain it. Unlike animals, fungi digest material outside the body and then absorb the digested material. They also contain chitin within their cell walls, unlike plants, and have a filamentous body plan. The filaments are called **hyphae**. Fungi also produce spores for reproduction. Fungi play important ecological roles, such as decomposition, and are a source of food for many animals including humans. Today you will examine various representatives of the kingdom Fungi.

PHYLUM ZYGOMYCOTA

Phylum Zygomycota should be familiar to most people as it contains bread mold that contaminates bread and other bakery products. Figure 15.1 shows the life cycle and structure of bread mold. **Sporangia** forms on the tips of the **sporangiophore**, which are aerial hyphae. The sporangia produce **spores**, which disperse and germinate when they land on a suitable food source. Obtain a slide of bread mold and observe the sporangia and sporangiophores. Use the space below to draw what you see.

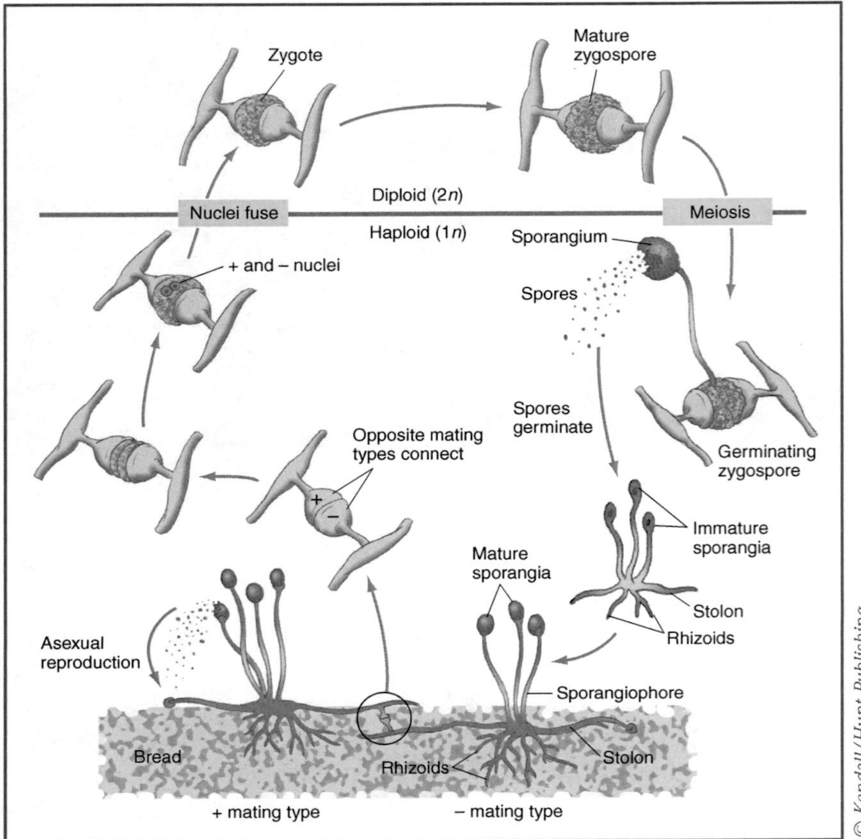

FIGURE 15.1

PHYLUM ASCOMYCOTA

Members of this phylum are the sac fungi. This group includes the truffles (Figure 15.2) and morels (Figure 15.3). This phylum gets its name because a sac called an **ascus** contains the spores (**ascospores**). A mature ascus contains eight ascospores. Ascospores release from the ascus during asexual reproduction and germinate in the soil. Figure 15.4 shows the life cycle of a typical member of this phylum. Obtain a slide of the genus *Peziza* and examine using the microscope. Use the space below to draw what you see. Pay close attention to the number of ascospores in each ascus.

FIGURE 15.2

FIGURE 15.3

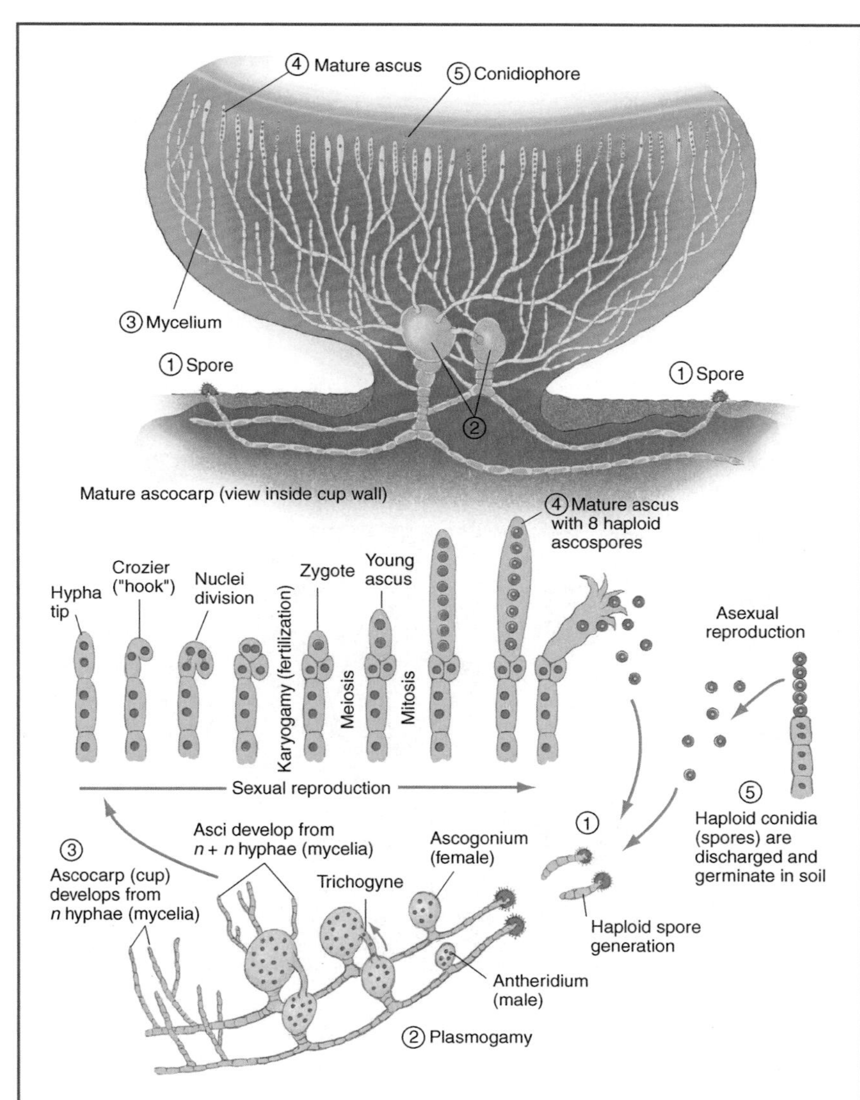

FIGURE 15.4

PHYLUM BASIDIOMYCOTA

This is the most familiar of all the fungal phyla as this contains the mushrooms (Figure 15.5). This phylum gets its name because a club-shaped structure called a **basidium** contains the spores (**basidiospores**). Figure 15.6 shows the life cycle of a mushroom. Located on the underside of the cap of the mushroom are slits called **gills** (Figure 15.7). This is where the basidia are located. Each basidium contains four basidiospores at maturity. Obtain a slide of the genus *Coprinus* and examine it using the microscope. Use the space below to draw what you see and compare it with Figure 15.8.

FIGURE 15.5

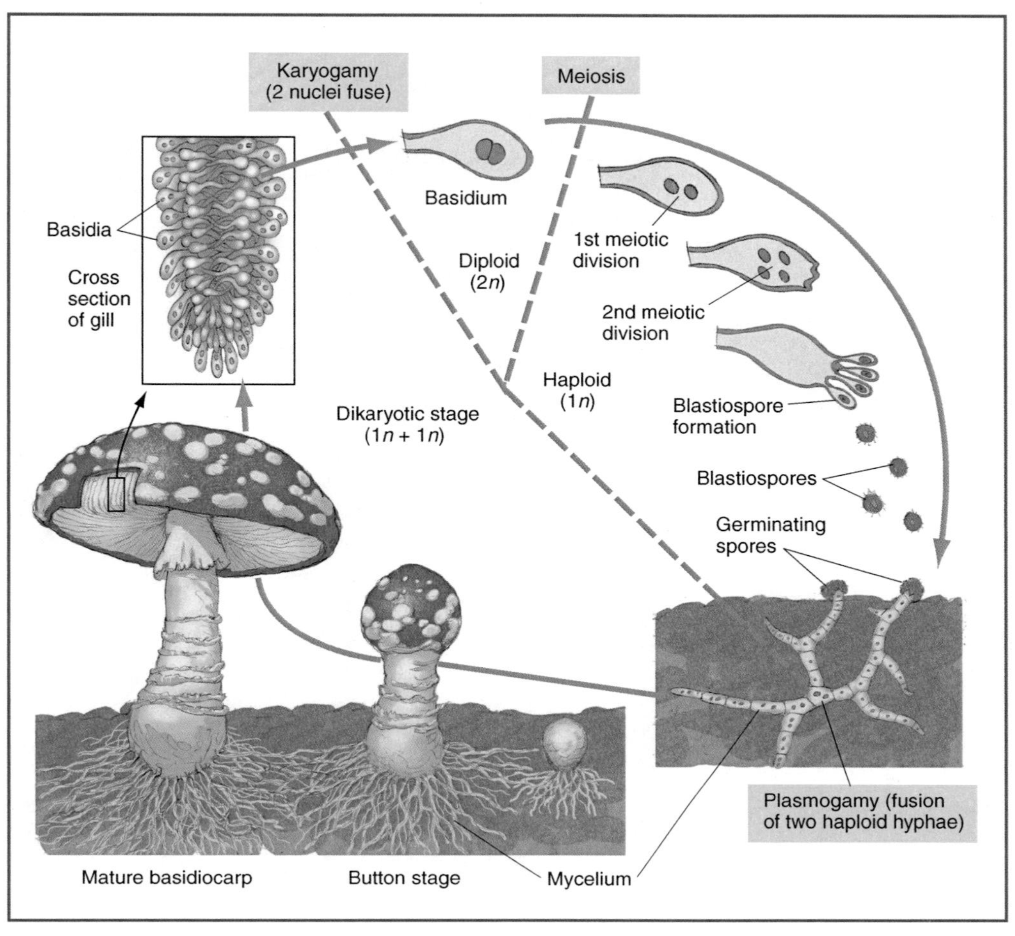

FIGURE 15.6

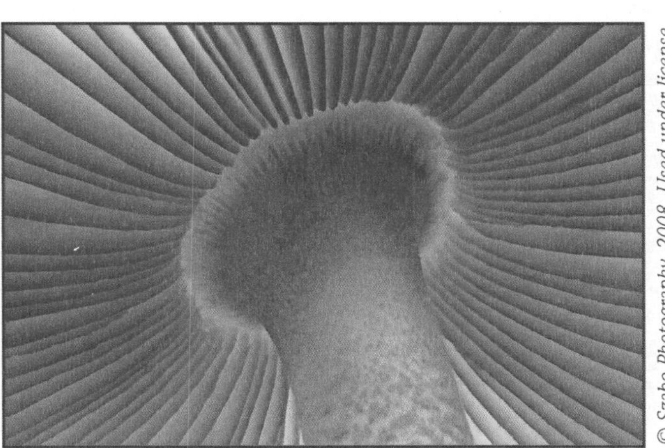

FIGURE 15.7

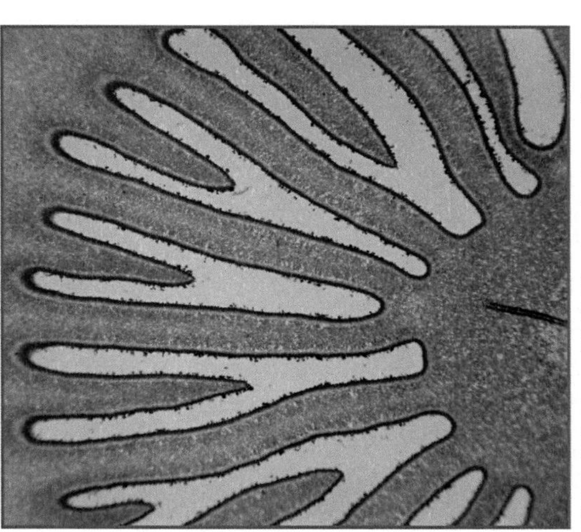

FIGURE 15.8

NAME ____________________

EXERCISE 15

QUESTIONS

1. What structure produces spores in Phylum Zygomycota?

2. Which phylum produces basidiospores?

3. In your own words describe the life cycle of a mushroom.

4. What ecological roles do fungi play in the environment? Use your textbook for help.

EXERCISE 16

PLANT ORGANIZATION

INTRODUCTION

If you view a photograph of the world taken from space and focus on the land masses you will notice one obvious feature: They are green. This is because for the most part plants dominate the terrestrial habitats. The plant kingdom is a very diverse group that has colonized every continent. Plants even dominated Antarctica once, although today they are no longer there. It is only fitting then that we should devote this lab and the next lab exercise to this important group that has allowed animals to flourish. Today we will focus on the structure of plants.

PLANT BODIES

Plant bodies divide into two general parts: the **root** and the **shoot**. Roots typically are used for absorption. However, there are modifications along this theme, and they are useful for other functions. Stems and leaves comprise the shoots. **Stems** can be woody or non-woody (**herbaceous**). **Leaves** are the photosynthetic structure of the plant. As might be expected there are specialized stems and leaves also. Take a look at Figure 16.1, which shows a generalized vascular plant body. Using your textbook as a reference define or give the function of the following structures:

Root cap

Root hairs

Vascular bundles

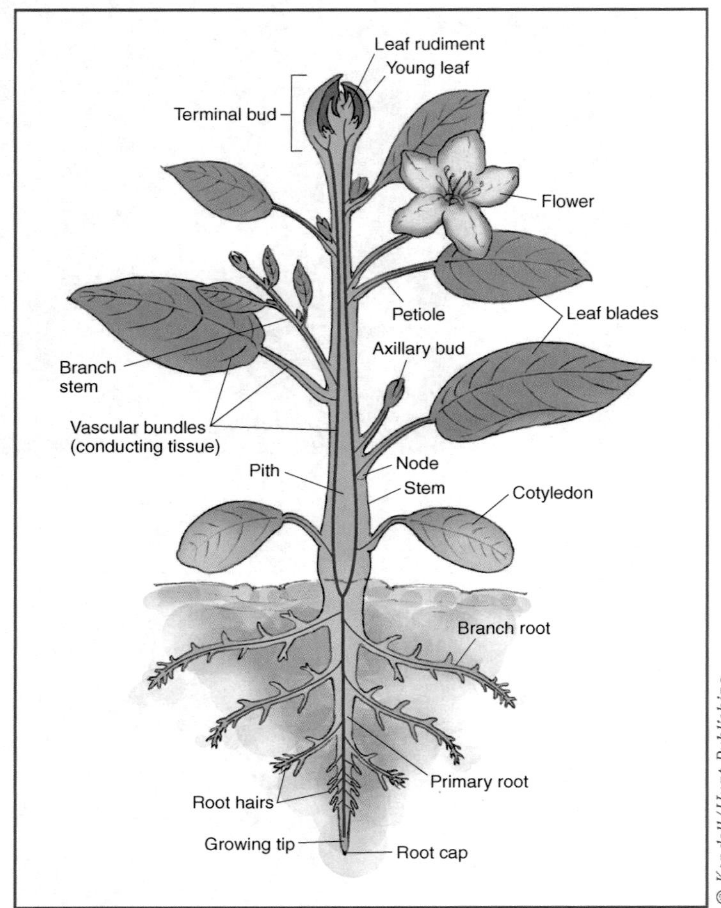

FIGURE 16.1

Cotyledon

Stem

Node

Pith

Petiole

Axillary bud

Terminal bud

Flower

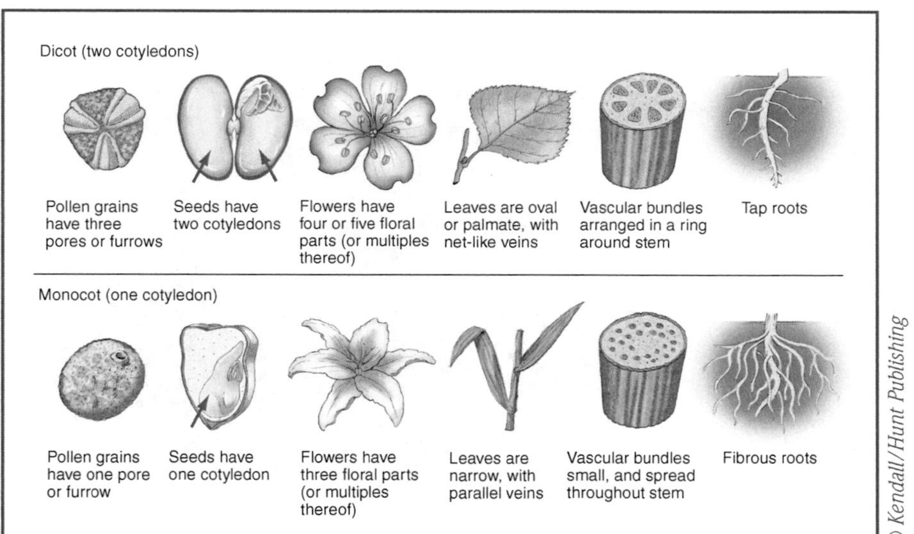

FIGURE 16.2

MONOCOTS AND EUDICOTS (DICOT)

Flowering plants divide into two groups: **monocots** and **eudicots** (**dicots**). Figure 16.2 shows the differences between these two groups. Using this figure as a guide, determine if the plants provided to you by your lab instructor are monocots or dicots.

Plant 1

Plant 2

Plant 3

Plant 4

Plant 5

LEAF STRUCTURE

Leaves are the photosynthetic structure of the plant. Essentially they are the mouths of the plant. Examine Figure 16.3, which shows a generalized leaf structure. Use your textbook as a reference to define or to give the function of the following structures:

Guard cells

Stoma

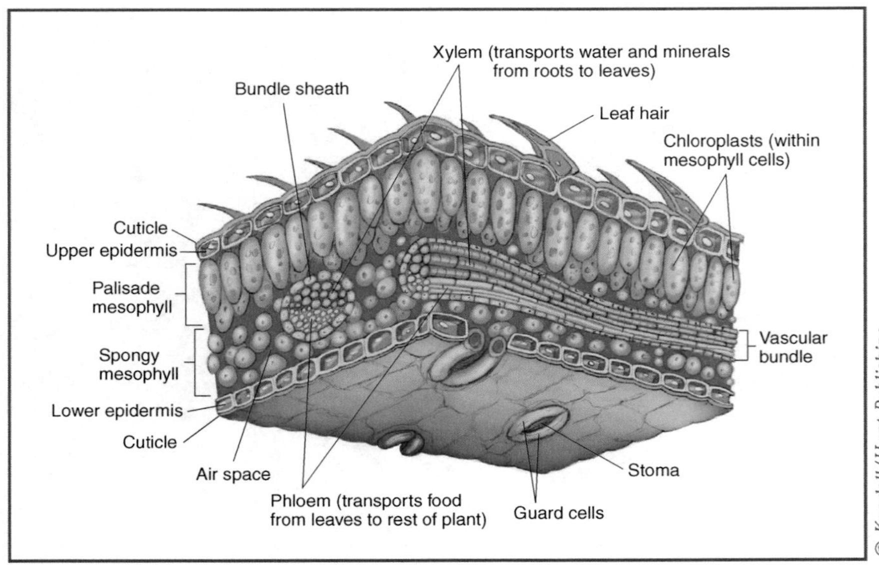

FIGURE 16.3

Cuticle

Spongy mesophyll

Palisade mesophyll

Bundle sheath

Xylem

Phloem

Leaf hair

Chloroplast

Vascular bundle

Obtain a cross sectional slide of a leaf and examine this structure with a microscope. Locate all possible structures listed in Figure 16.3 and use the space below to draw what you see.

ARRANGEMENTS OF LEAVES

Leaves can become arranged in a variety of configurations depending on the species. It is important to have a basic familiarity of leaf arrangements in order to properly identify plant species. Figure 16.4 shows the basic arrangements of leaves. Use this figure as a guide to determine the leaf arrangement of leaves provided to you by your instructor.

Plant 1

Plant 2

Plant 3

Plant 4

Plant 5

FIGURE 16.4

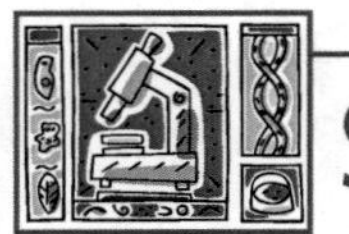

STEMS

The structure of stems will differ between monocots and dicots. Additionally, within the dicots there are herbaceous and woody varieties. Examine Figure 16.5. This figure shows the cross section of an herbaceous monocot stem. Notice that the **vascular bundles** are scattered throughout the stem. This is a characteristic of monocots. Obtain a slide of the monocot *Zea mays* and examine this using the microscope. Locate all structures listed in Figure 16.5. Use the space below to draw and label what you see.

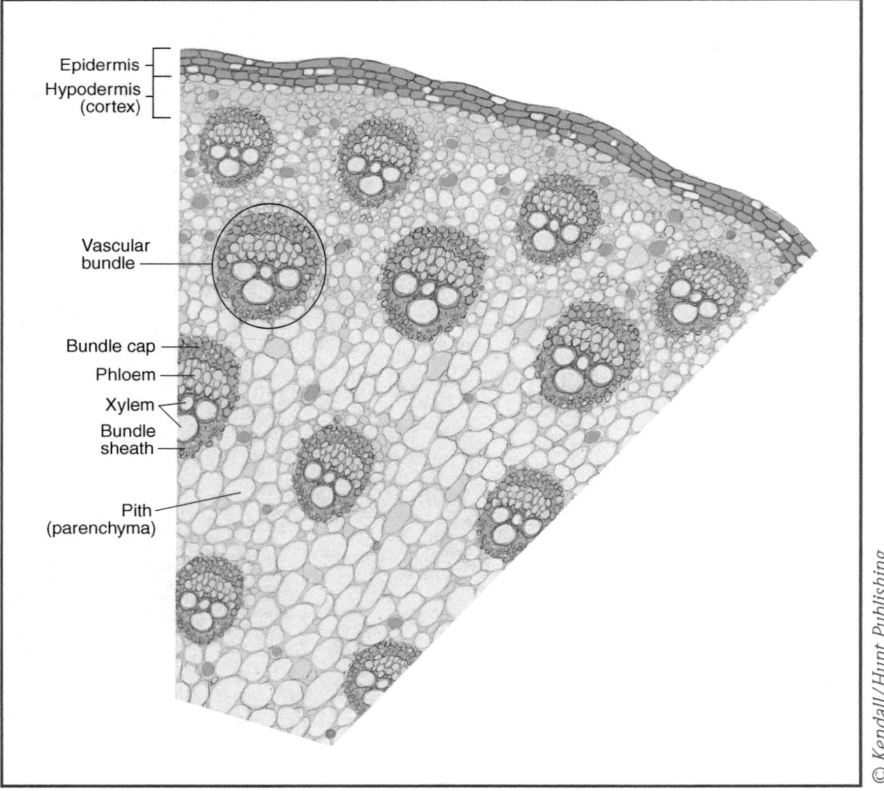

FIGURE 16.5

Examine Figure 16.6. This figure shows the cross section of an herbaceous dicot stem. Notice that the vascular bundles are in a ring around the edge of the stem. This is a characteristic of dicots. Obtain a slide of the herbaceous dicot *Cucurbita* and examine this using the microscope. Locate all structures listed in Figure 16.6. Use the space below to draw and label what you see.

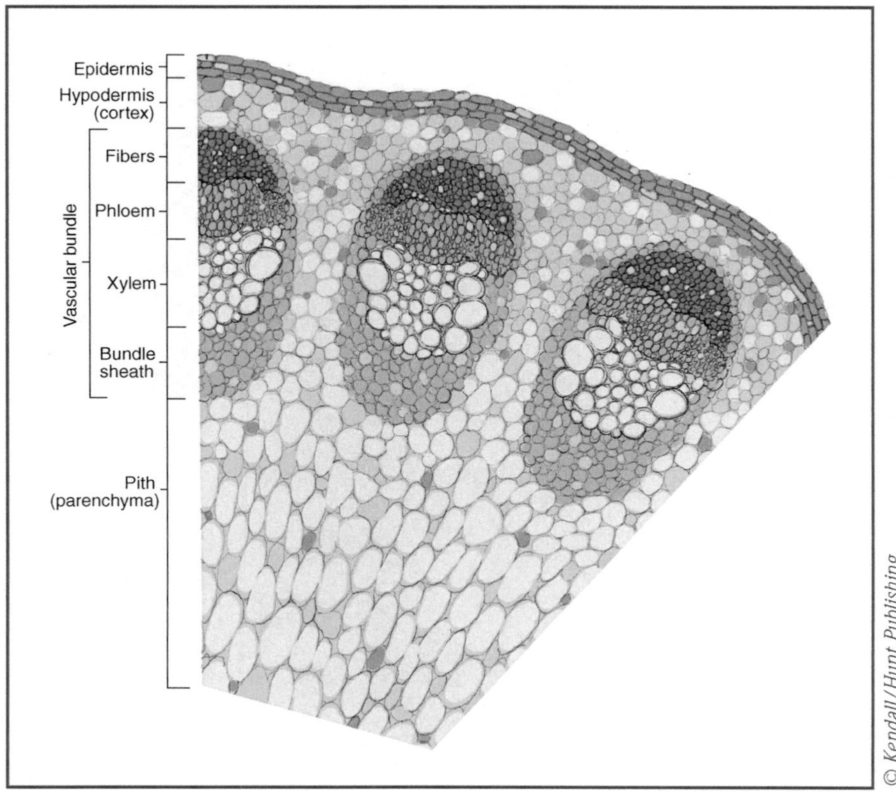

FIGURE 16.6

Examine Figure 16.7. This figure shows the cross section of a woody dicot stem. Notice the annual rings. These rings form at the creation of **xylem** each year. Also notice that the vascular bundles arrange in a ring like in the herbaceous dicot. However, the arrangement of the xylem and **phloem** is different between the two dicots. Obtain a slide of the woody dicot *Tilia* and examine this using the microscope. Locate all structures listed in Figure 16.7. Use the space below to draw and label what you see.

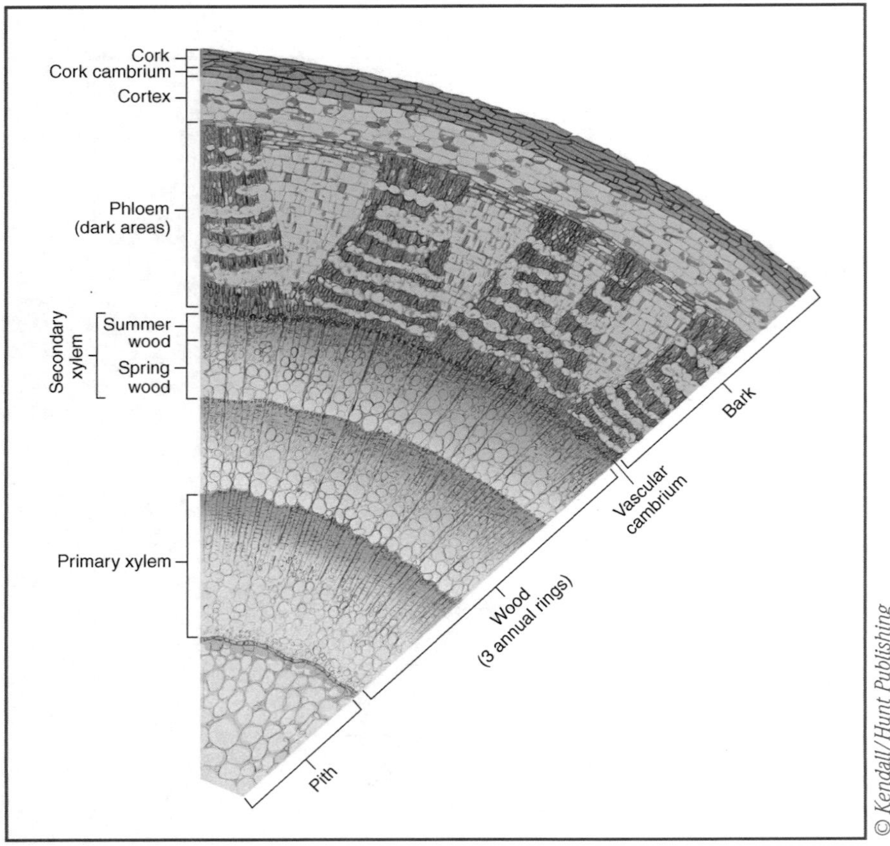

FIGURE 16.7

ROOTS

Roots primarily absorb water and nutrients from the soil and transport those materials to the shoots. Figure 16.8 shows a generalized root structure as well as cross sections of the roots of monocots and dicots. Obtain a monocot root slide and a dicot root slide and examine those using the microscope. For each slide locate the structures listed in their representative cross sections in Figure 16.8. Use the space below to draw what you see.

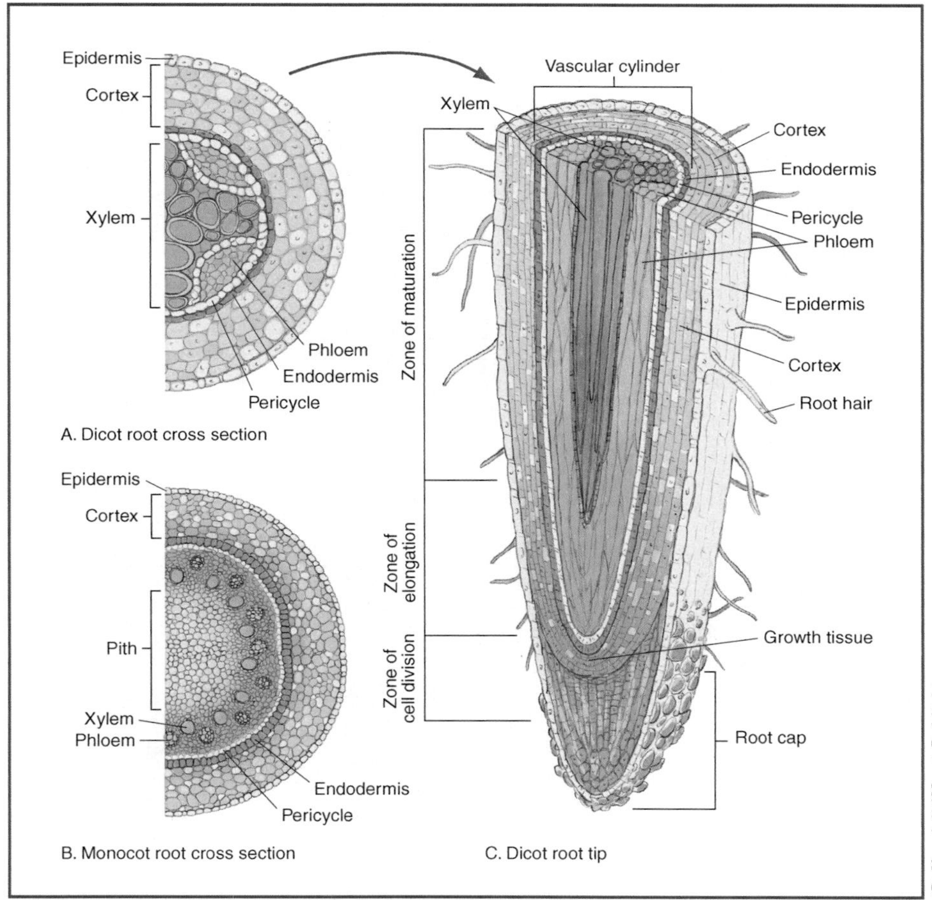

FIGURE 16.8

NAME ____

EXERCISE 16

QUESTIONS

1. Draw an herbaceous dicot stem and label the parts.

2. Draw a woody dicot stem and label the parts.

3. What are the differences between leaves and roots of monocots and dicots?

EXERCISE 17

REPRODUCTION IN THE FLOWERING PLANTS

INTRODUCTION

Plants have a variety of ways to reproduce; however, most people are familiar with the **Angiosperms**, which reproduce utilizing **flowers**. Flowering plants are very important to humans as most of our agricultural crops come from flowering plants. Even honey, which many people enjoy, relies upon flowers and their **pollinators**. Today we will examine reproduction in flowering plants.

FLOWERS

Flowers are the reproductive structure of plants, and they come in many varieties and sizes from the smallest flower in the world, duckweed (*Lemnaceae,* Figure 17.1) to the largest flower in the world *Rafflesia* (Figure 17.2). Its flower can grow to 100 cm in diameter and weigh as much as 10 kg. Regardless of their size, all flowers have the same basic structure, as shown in Figure 17.3. We call the female parts of

FIGURE 17.1

FIGURE 17.2

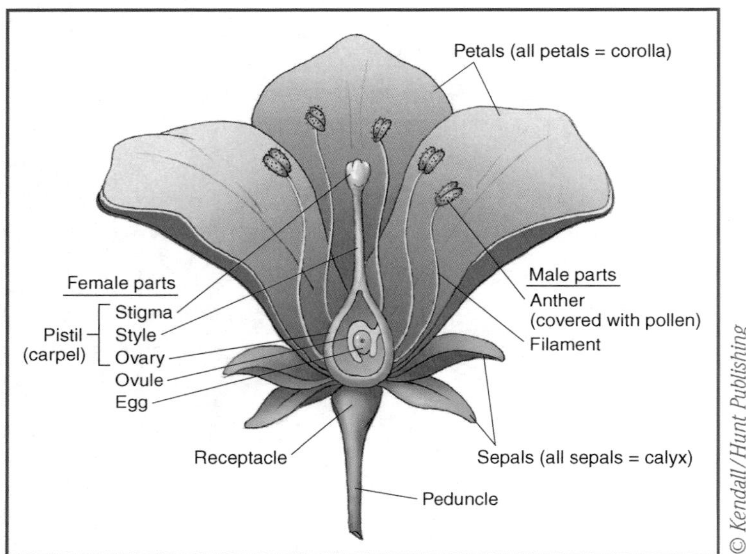

FIGURE 17.3

the flower collectively a **pistil** or **carpel** and the male parts collectively the **stamen.** Using your textbook as a reference define or list the function of the following structures:

Stigma

Style

Ovary

Ovule

Egg

Anther

Filament

Sepals

Petals

Receptacle

Peduncle

Obtain a *Lilium* flower and observe all parts possible as listed in Figure 17.3. Remove pollen from the anther and make a wet mount. Observe these structures with the aid of a microscope.

POLLINATION

The **petals** of the flower are usually conspicuous, which is by design. The whole purpose of the petals is to attract the attention of a pollinator, and they do so with the promise or the illusion of some reward. The co-evolution of flowers and their pollinators is one of the most extraordinary stories in all of biology. The color, markings, scent, and time of bloom are all designed to attract either a particular species of pollinator or a particular group of pollinators. Take a look at the situations below and try to determine what the pollinator is.

1. *Rafflesia* has a flower that smells like rotted meat, so much so that some have even called it the corpse flower. What do you think the pollinator is?
2. Flowers that open at night have what type of pollinator?

FIGURE 17.4

3. Take a look at Figure 17.4. This shows a bee orchid, which is an orchid that mimics a bee. As stated earlier, flowers attract pollinators by the promise or the illusion of a reward. The reward in this case is an illusion. The bee that comes to this plant is essentially fooled into believing it will get a fantastic reward, what do you think it is?
4. Many (but not all) flowers that have a red color also have no scent. What do you think their pollinators are? *Hint:* Why are hummingbird feeders red?

Regardless of the pollinator used, the goal is to transfer pollen from the **anther** of one flower to the **stigma** of another flower. The anther of the flower bears **pollen,** which contains sperm for fertilization. Examine Figure 17.5, which shows a cross section of an anther. Notice the pollen grains being formed. At maturity this structure will open and release the pollen. Obtain a slide of a *Lilium* anther and compare it to Figure 17.5. Use the space below to draw what you see.

When pollen lands on the stigma a **pollen tube** grows from the pollen grain (Figure 17.6). This tube grows down the length of the style and finally reaches the egg. Two sperm travel down the length of the pollen tube, and one fertilizes the egg, forming a **zygote**, while the other fuses with polar bodies and forms the **endosperm**. Called **double fertilization**, it demonstrates the angiosperm life cycle. See Figure 17.7.

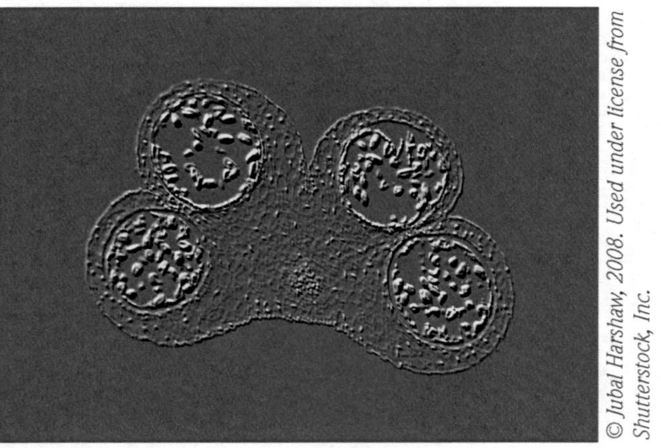

FIGURE 17.5

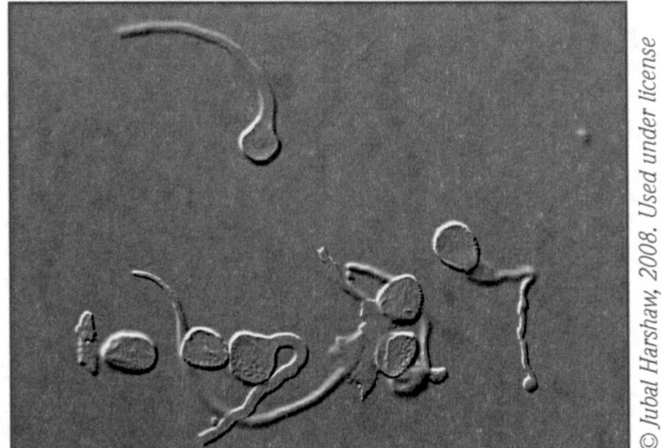

FIGURE 17.6

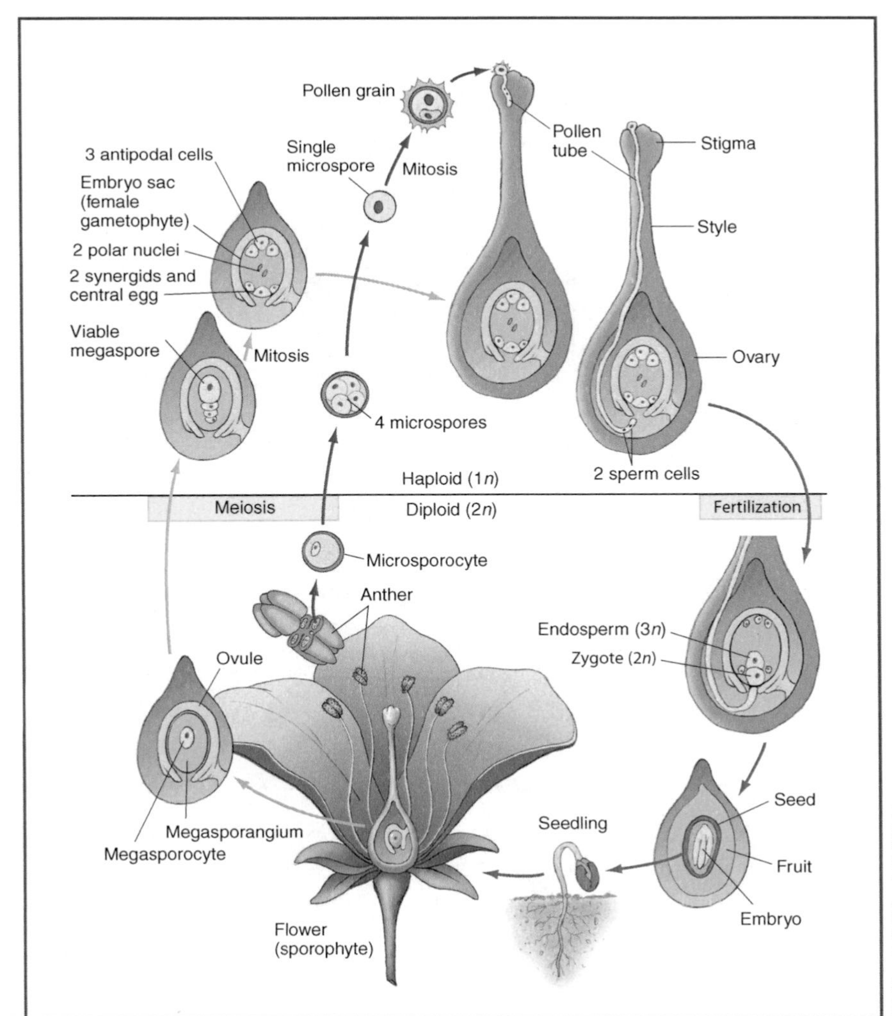

FIGURE 17.7

Answer the following questions regarding flowers:

1. What is the purpose of the anther?

2. What is the purpose of the style?

3. Why do some flowers smell good while others smell bad?

4. Why are flowers conspicuous?

5. In your own words summarize the life cycle of angiosperms as depicted in Figure 17.7.

FRUITS

Once **fertilization** has occurred a **seed** forms that houses the developing **embryo.** Plants need to have their seeds dispersed so they are not competing with their own offspring for the same resources. In order to accomplish this task fruits form. Fruits are any structure that contains seeds, and they typically form from the ovary or a group of ovaries of the plant (Figure 17.8). Fruits come in a variety of types from dry fruits to fleshy fruits as shown in Figure 17.9. Whatever its ultimate form, the purpose of a fruit is seed dispersal. Plants use a variety of methods to disperse their seeds, including animals, wind, water, and explosive dehiscence. Examine the fruits provided to you by your instructor and determine what type of fruit it is and what you think the method of dispersal used by that fruit is.

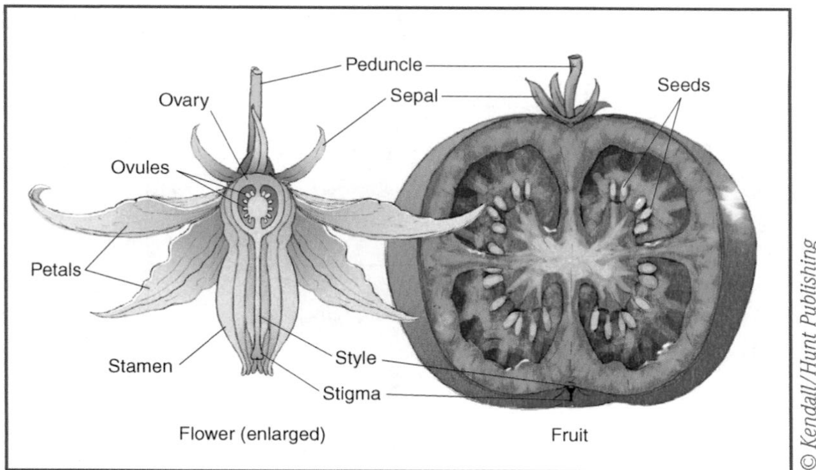

FIGURE 17.8

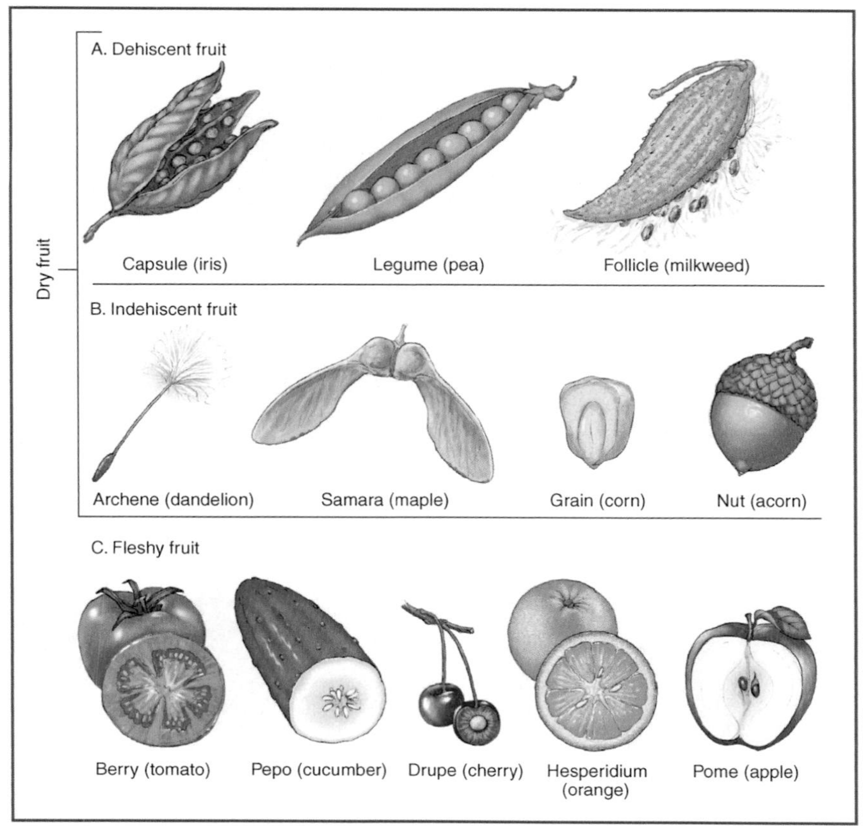

FIGURE 17.9

FRUIT (NAME)	TYPE	DISPERSAL METHOD
1		
2		
3		
4		
5		
6		
7		
8		

Answer the following questions regarding fruits:

1. Why would a plant produce a brightly colored fruit that tastes sweet? Do you really think a plant would want its seeds eaten?

2. An oak tree might make thousands of acorns in a single season. Why? If a tree can develop from a single acorn, why not save the energy and just make one?

3. What do you think would happen to a developing plant that did not disperse from its parent?

NAME ______________________________

EXERCISE 17

QUESTIONS

1. Draw and label the female parts of a flower.

2. Draw and label the male parts of a flower.

3. Define or list the function of the following.

Stigma

Style

Ovary

Ovule

Egg

EXERCISE 18

ANIMALS: THE INVERTEBRATES

INTRODUCTION

Kingdom Animalia is a very diverse group that includes humans (*Homo sapiens*), among others, but what makes an animal? Animals have several characteristics in common. They are multicellular **eukaryotes** and are **heterotrophic**; however, they must first ingest their food before digestion, unlike the fungi. Typically animals are mobile at least during some stage of their development. Animals are capable of sexual reproduction, and their embryo undergoes developmental stages. Despite these features in common, animals have evolved ingenious methods to survive within their respective habitats. Animals are so successful that they have colonized all the world's oceans and continents, including some of the harshest environments on earth. We will begin our exploration of this kingdom with the invertebrate phyla.

PHYLUM PORIFERA

Phylum Porifera includes the **sponges** (Figures 18.1 and 18.2). Sponges are typically marine animals. Their body shape is like a sack or a tube, and it contains pores. Figure 18.3 shows the typical body plan of a sponge. They are multicellular like all animals, but their bodies lack organized **tissues**. Sponges are grouped based on the composition of their skeleton. Some have **spicules** within their skeletons (these are needle-like structures) while others have skeletons made of a protein called **spongin**, which is a soft protein. Sponges are **sessile** (non-mobile) filter feeders as adults. (Only the zygote is capable of movement.) Water moves through the pores in the body of the sponge, and **collar cells** in the body filter microscopic food particles and digest them.

FIGURE 18.1

FIGURE 18.2

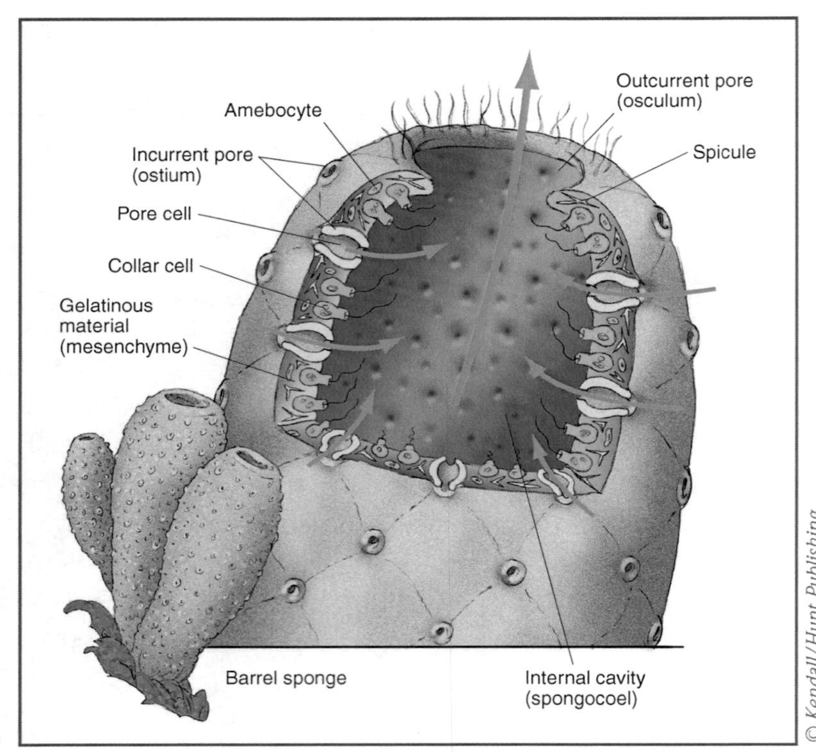

FIGURE 18.3

Obtain a prepared slide of sponge spicules and observe them with the aid of a microscope. Draw what you see in the space provided.

Obtain a cross-section slide of *Grantia* and locate the pores, the osculum, amoebocytes, and the collar cells. Use the space below to draw what you see.

Your lab instructor has placed several sponges for you to observe. Are their bodies made of spicules or spongin?

PHYLUM CNIDARIA

Phylum Cnidaria contains the **jellyfish** (Figure 18.4), **sea anemones** (Figure 18.5), and coral. Cnidarians are the first animal phyla with true tissues, and they have two basic body forms: the **polyp** and the **medusa.** The polyp is sessile while the medusa is the mobile form. In many Cnidarians one form is the dominant body plan. For instance, the dominant body plan for a jellyfish is the medusa; however, the sea anemone's dominant form is the polyp. Regardless of which form is dominant they both contain **tentacles** populated with **cnidocytes**, which are specialized cells designed for stinging which are unique to this phylum. Within the cnidocytes are **nematocysts**, which contain a threadlike fiber used to subdue and capture prey. Many of the nematocytes contain toxins to kill prey. Once something touches the cnidocyte it will fire a nematocyte which injects poison in the victim. Some of these organisms can be quite harmful to humans such as the **Portuguese man-of-war** and the **Australian box jelly** which is so toxic that it can kill a human. Despite these dangers some animals, such as clownfish, have evolved mechanisms to deal with the toxins and use Cnidarians for their home.

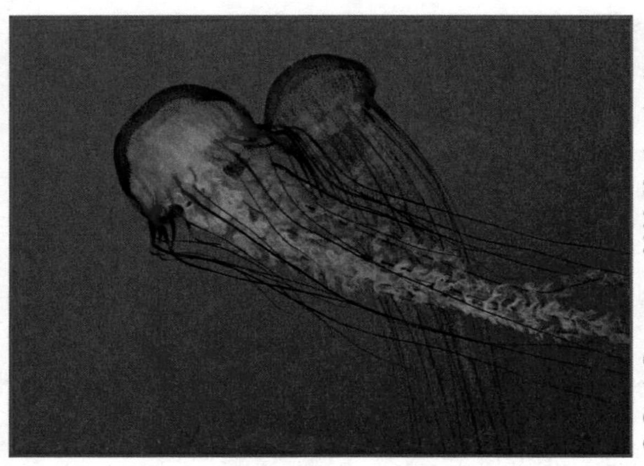

FIGURE 18.4

FIGURE 18.5

Obtain a prepared slide of the freshwater Cnidarian *Hydra* and examine it using the microscope. Notice the tentacles and the mouth. Use the space below to draw what you see.

PHYLUM PLATYHELMINTHES

This phylum contains the **flatworms**, which include the planarians, tapeworms, and flukes. The last two are parasites. Flatworms have organ systems, such as muscles, digestive, excretory, and reproductive systems. However, no respiratory system exists in this group. Because they are flat and thin, gas exchange can occur via diffusion. This is an important group specifically because of the parasites that this phylum contains. Tapeworms can infest humans that eat undercooked meat contaminated with their cysts. They grow in the human host and rob it of nutrients. Flukes are also parasites of vertebrates, including humans. The blood fluke, for instance, infects humans and is responsible for **schistosomiasis**, which kills over a half million people each year in the Middle East, Asia, and Africa.

Examine Figure 18.6. This figure shows the internal structures of the planarian. Obtain a prepared slide of the planarian and examine this with the aid of a

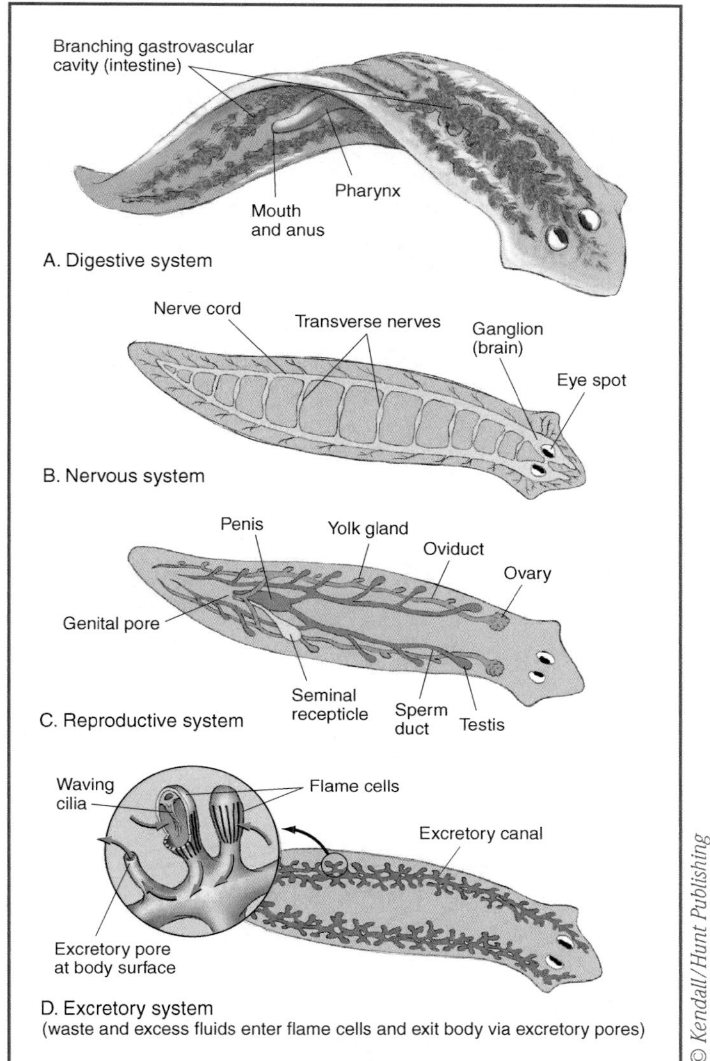

FIGURE 18.6

microscope. Compare this slide with Figure 18.6 and identify all structures visible. Use the space below to draw what you see.

PHYLUM NEMATODA

This phylum contains the **roundworms**. Roundworms are typically described as a "tube within a tube," as demonstrated in Figure 18.7. This group contains both beneficial species and harmful species, including parasites. Obtain a slide of the genus *Ascaris* and examine it with the aid of a microscope. Compare the slide with Figure 18.7 and identify all structures visible. Use the space below to draw what you see.

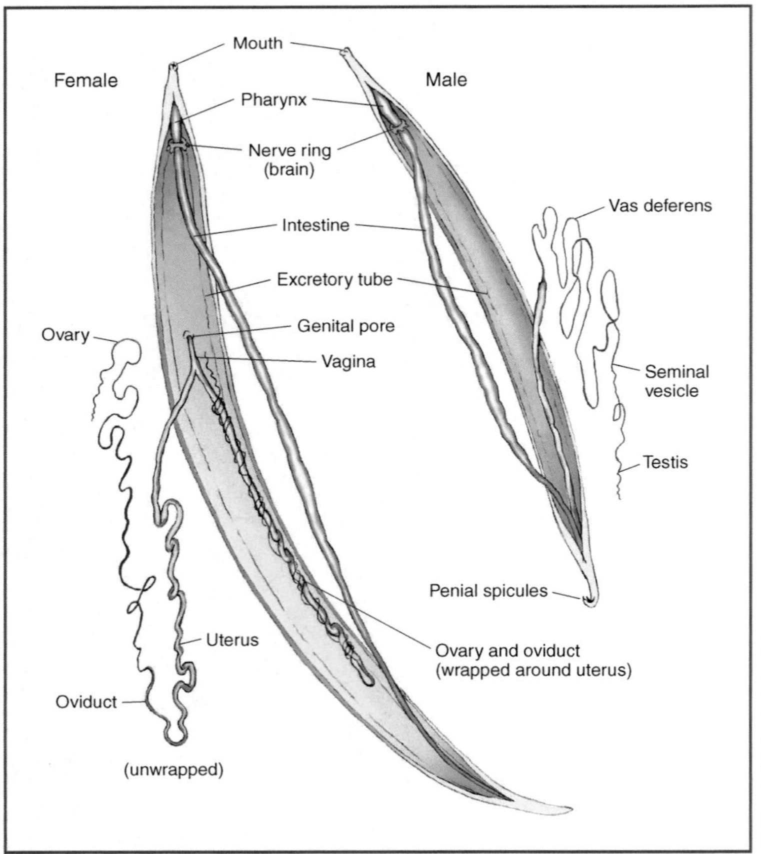

FIGURE 18.7

PHYLUM MOLLUSCA

This is a diverse group that contains snails, clams, mussels, oysters, scallops, octopi, and squid. Despite the diversity within this group they all share several features. These include a **visceral mass** containing internal organs, a **foot** used for movement, and a **mantle** for protection. In some species the mantle secretes a **shell** for protection. This group is responsible for the sea shells that beachgoers spend time collecting.

Examine Figure 18.8 and familiarize yourself with the internal structure of a typical mollusc: the clam. Compare this figure with models of clams or observe a dissected specimen and locate all structures.

Figure 18.9 shows the external and internal structure of the squid. Obtain a dissecting tray, scissors and/or scalpel, gloves, and a preserved squid for examination. Gently cut along the side margin of the squid to expose the internal organs beneath and identify all structures within.

PHYLUM ANNELIDA

This phylum comprises the **segmented worms**, such as earthworms and leeches. This group's segmentation is evident when examining the rings located on the outside of the body of an earthworm. This group is more complex than the roundworms in that there is a more complex nervous system, including a **ventral nerve chord**, a more specialized digestive tract, and a **closed circulatory system** with blood vessels.

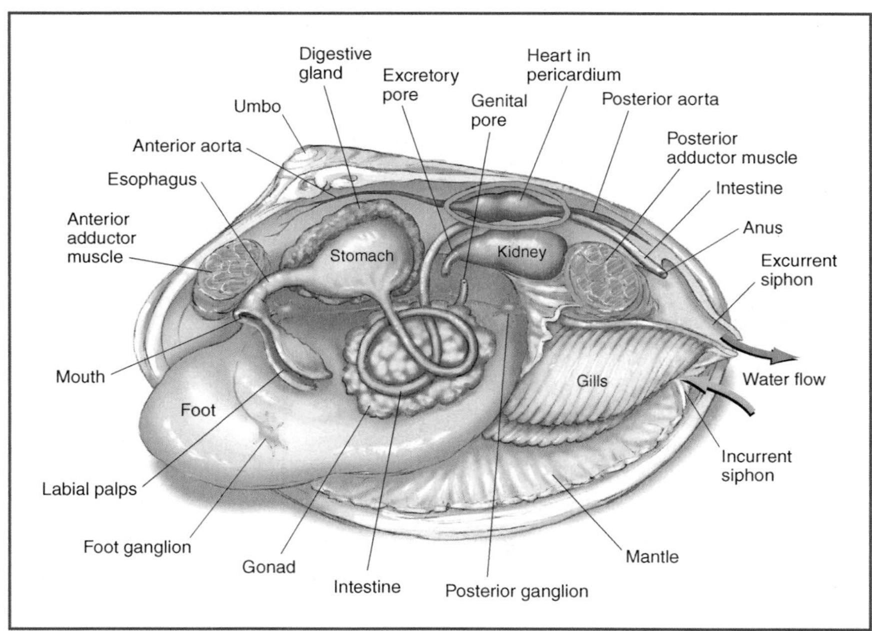

FIGURE 18.8

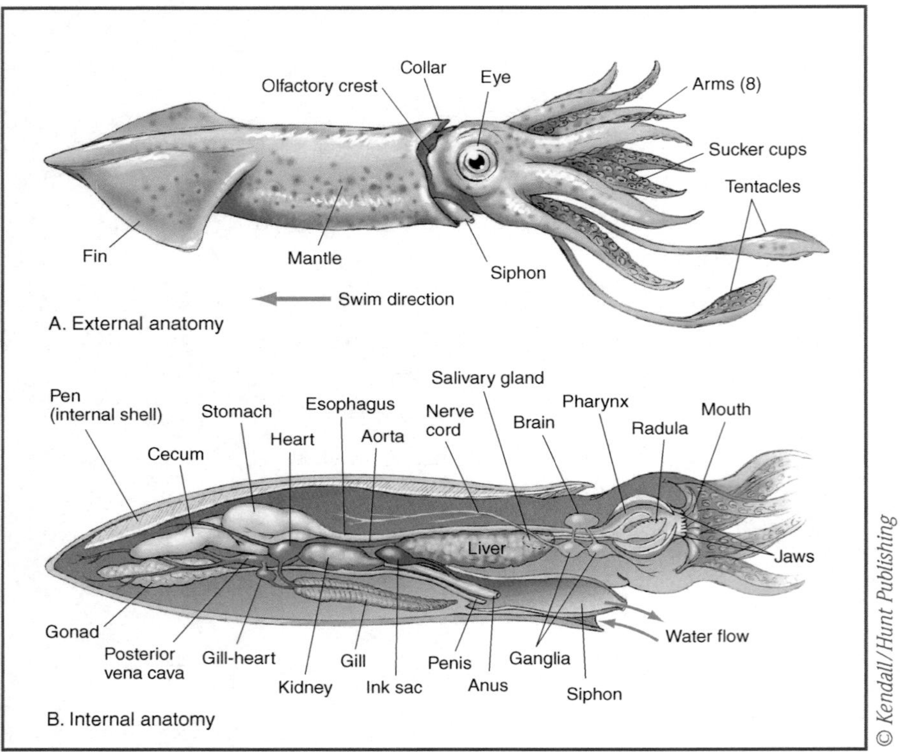

FIGURE 18.9

Obtain a prepared slide of the earthworm in cross section and examine it with the aid of a microscope. Compare this slide with Figure 18.10. Locate and identify all structures listed on Figure 18.10 visible on the slide. Use the space below to draw what you see.

Obtain a tray, scissors and/or scalpel, gloves, and an earthworm for dissection. Gently cut through the outer surface of the earthworm as depicted in Figure 18.11 and pin it open for examination. Locate and identify structures shown in Figures 18.12 and 18.13.

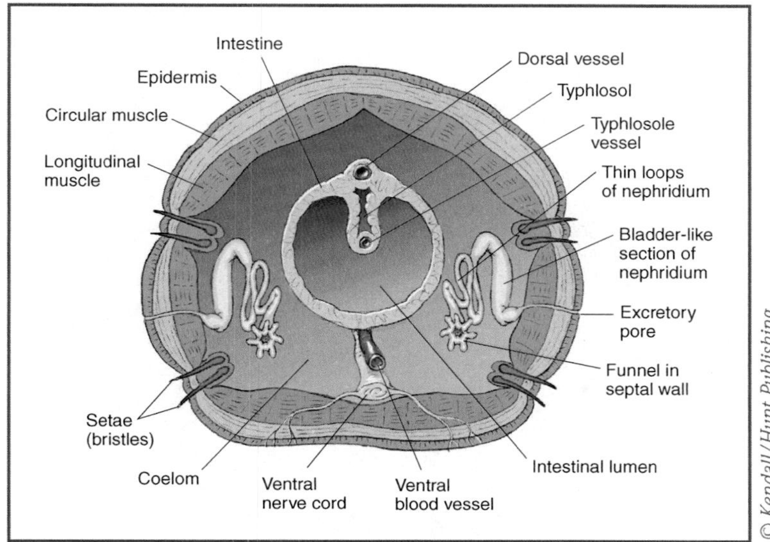

FIGURE 18.10

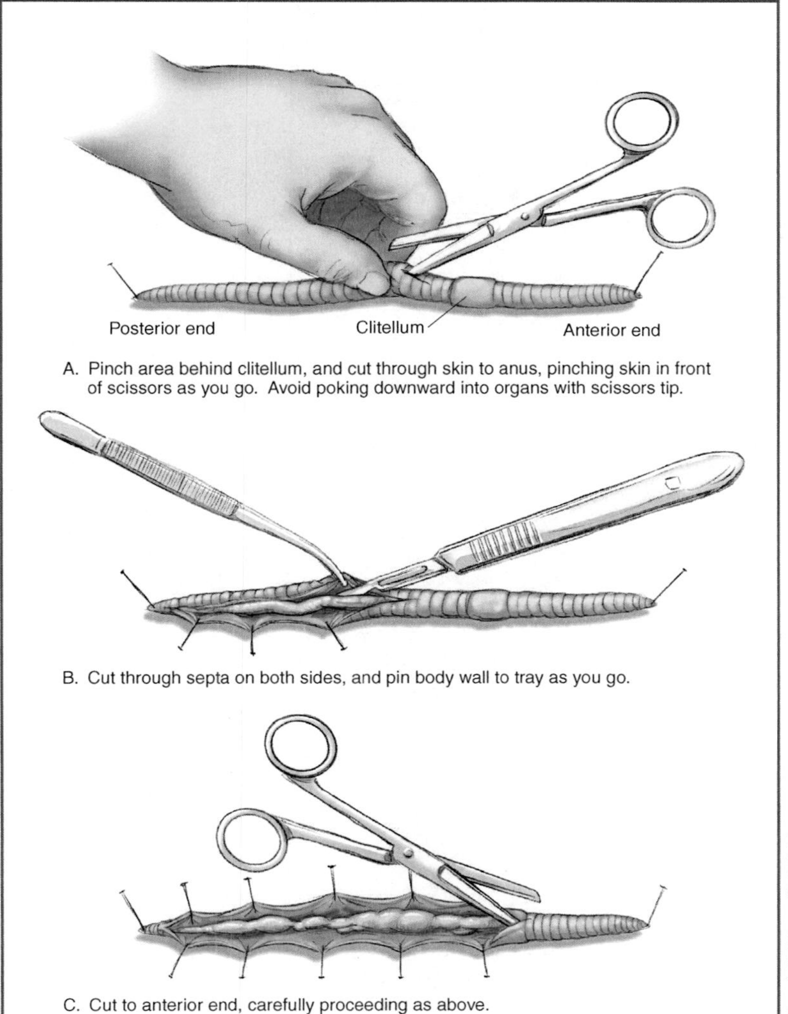

FIGURE 18.11

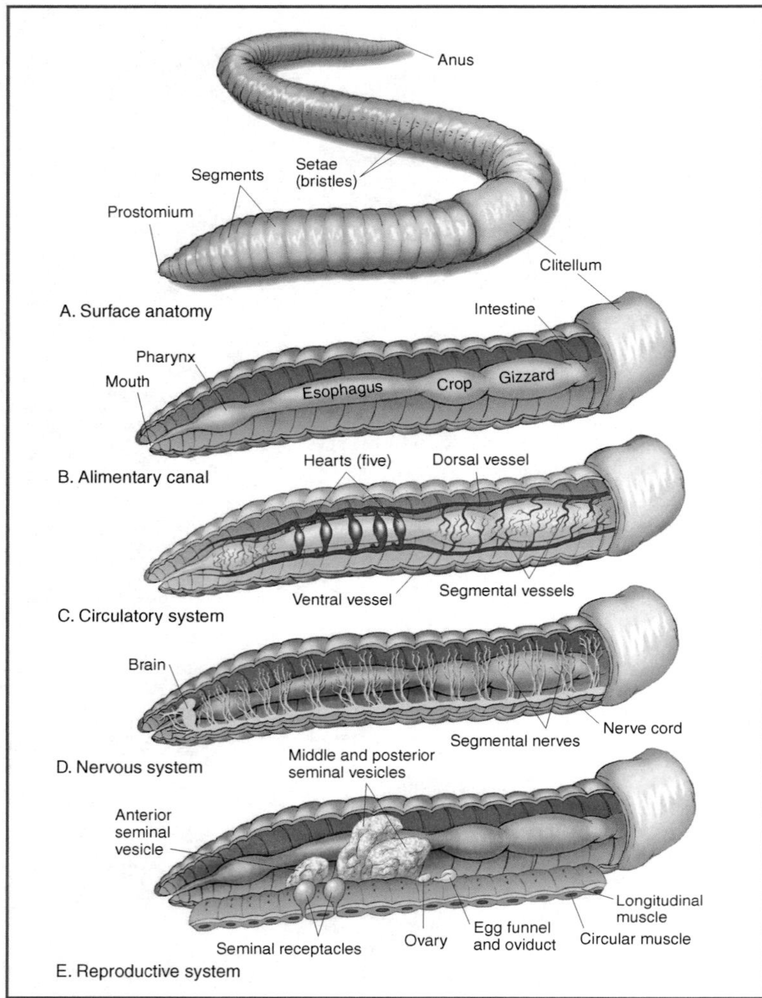

FIGURE 18.12

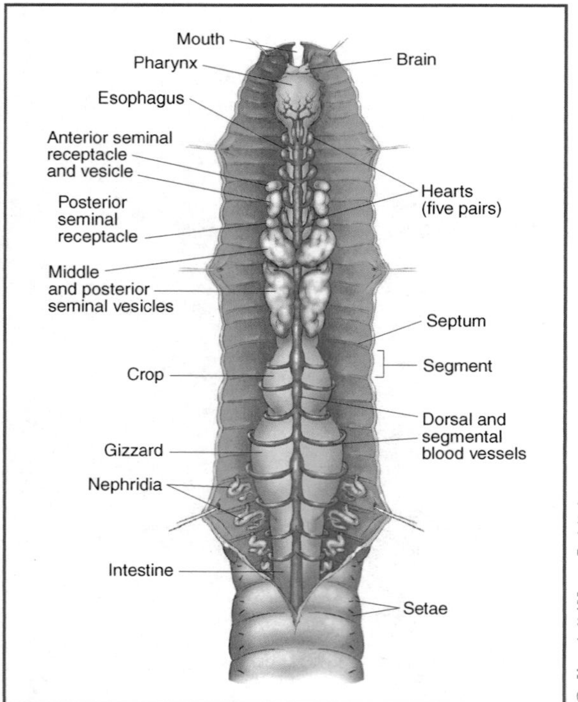

FIGURE 18.13

PHYLUM ARTHROPODA

This is one of the most diverse groups of animals you will study. So far there are over one million species known to science. This group contains such familiar species as insects, spiders, scorpions, crabs, lobsters, centipedes, millipedes, and others. The defining characteristic of this group is their **jointed appendages.** They also contain a well-developed nervous system, which includes a **ventral nerve chord** and a **brain.** A simple lab exercise cannot do this group justice as you can spend a lifetime studying this phylum, but in the interests of time we will focus on a couple of representative specimens.

Obtain a preserved crayfish and compare the external and internal anatomy to Figure 18.14.

Obtain a preserved grasshopper and compare the external and internal anatomy to Figure 18.15.

Using Figure 18.15 as a guide to label the structures indicated in Figure 18.16.

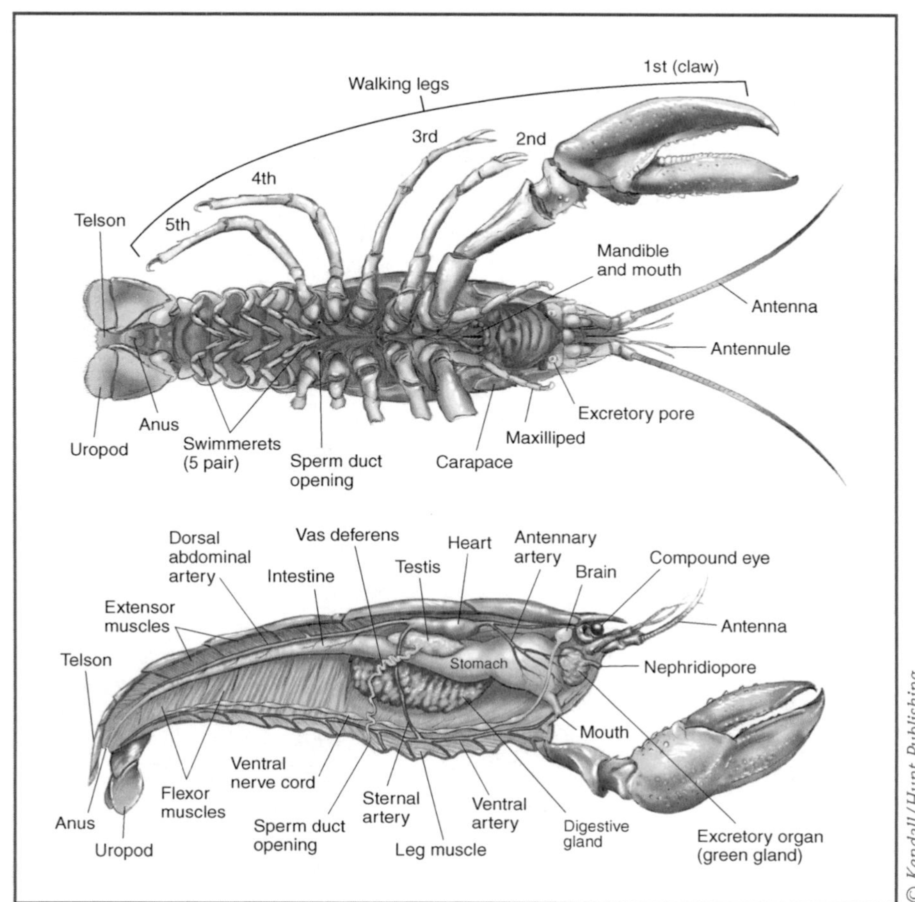

FIGURE 18.14

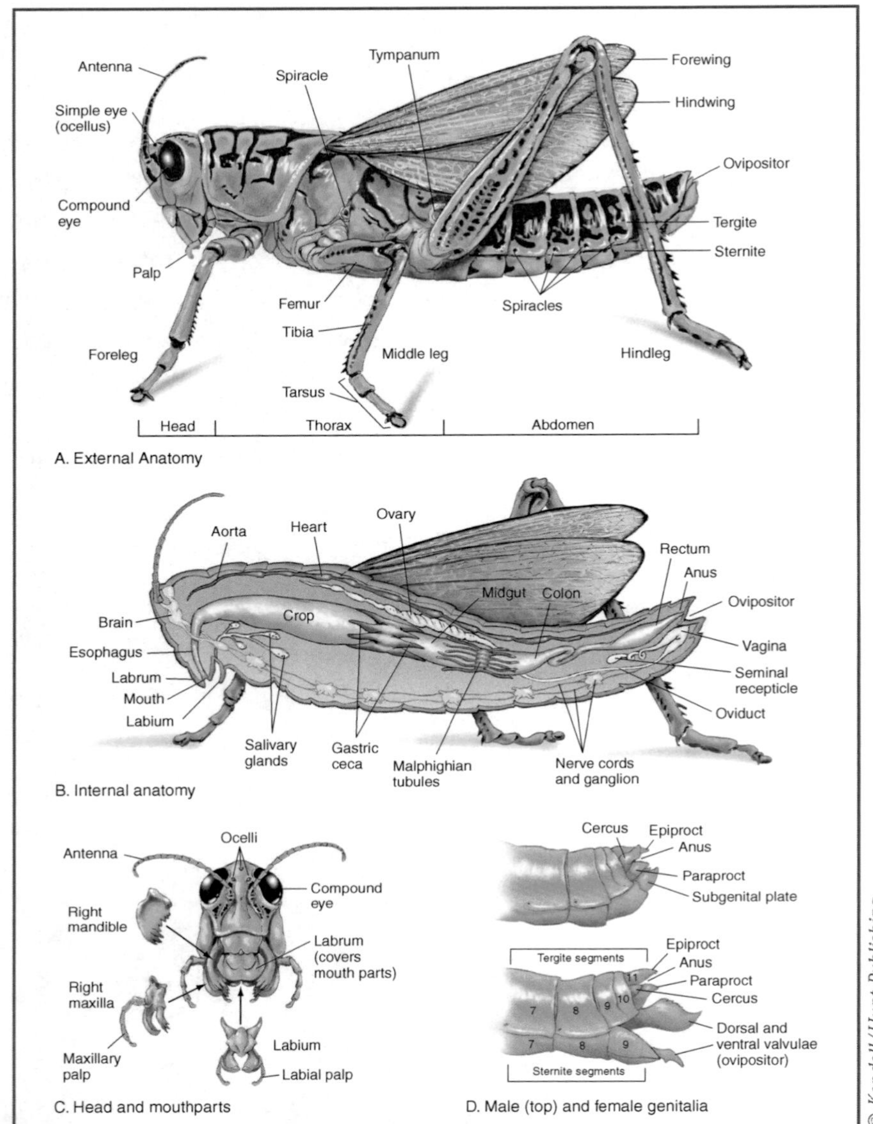

FIGURE 18.15

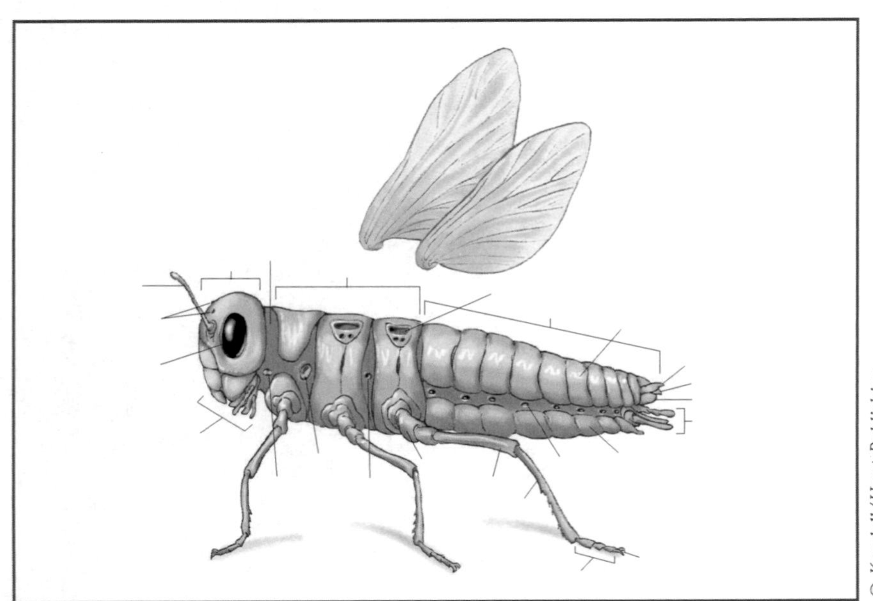

FIGURE 18.16

NAME ______________________________

EXERCISE 18

QUESTIONS

1. What are the characteristics of animals?

2. Define the following.

Nematocyst

Closed circulatory system

Visceral mass

Collar cells

3. List the phylum (or phyla) which contain the following.

   Mantle

   Cnidocyte

   Collar cells

   Ventral nerve chord

   Sponges

   Ticks

   Roundworms

   Earthworms

EXERCISE 19

PHYLA ECHINODERMATA AND CHORDATA

INTRODUCTION

As we continue in the Kingdom Animalia, we will now turn our attention to the final two animal phyla: Echinodermata and Chordata

PHYLUM ECHINODERMATA

Members of this group are marine and include the **sea stars** (Figure 19.1), **sea urchins** (Figure 19.2), **sea cucumbers** (Figure 19.3), **sea lilies** (Figure 19.4), and the **brittle stars** (Figure 19.5). Members of this group have **spines** on their outer covering, which are outgrowths of their internal skeleton. These spines give this

FIGURE 19.1

FIGURE 19.2

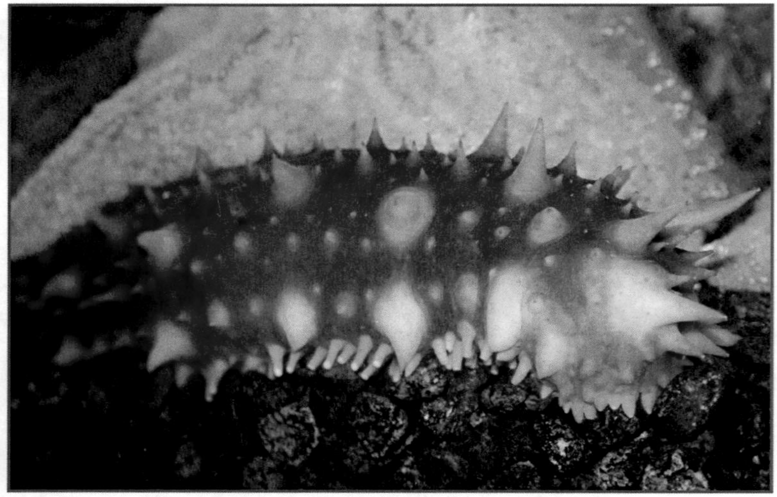

FIGURE 19.3

FIGURE 19.4

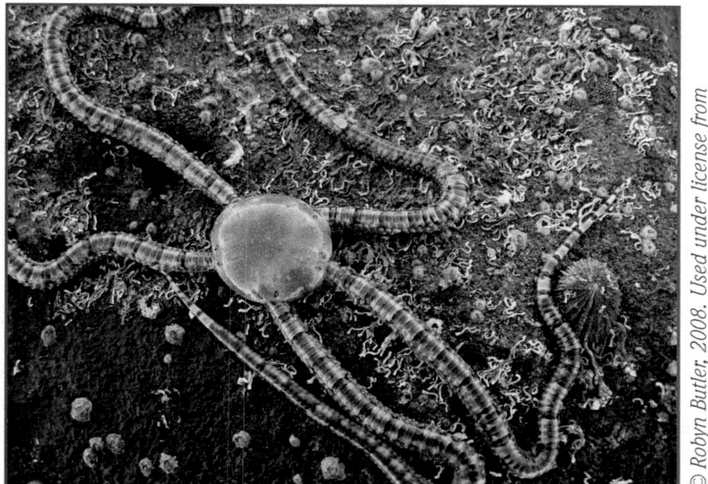

FIGURE 19.5

phylum its name. Echinoderms also contain a **water vascular system**, which aids in their movement. We will focus on a single representative species of this phylum: the sea star. Figure 19.6 shows the anatomy of the sea star (starfish). Familiarize yourself with this organism and compare it with preserved specimens provided to you by your instructor.

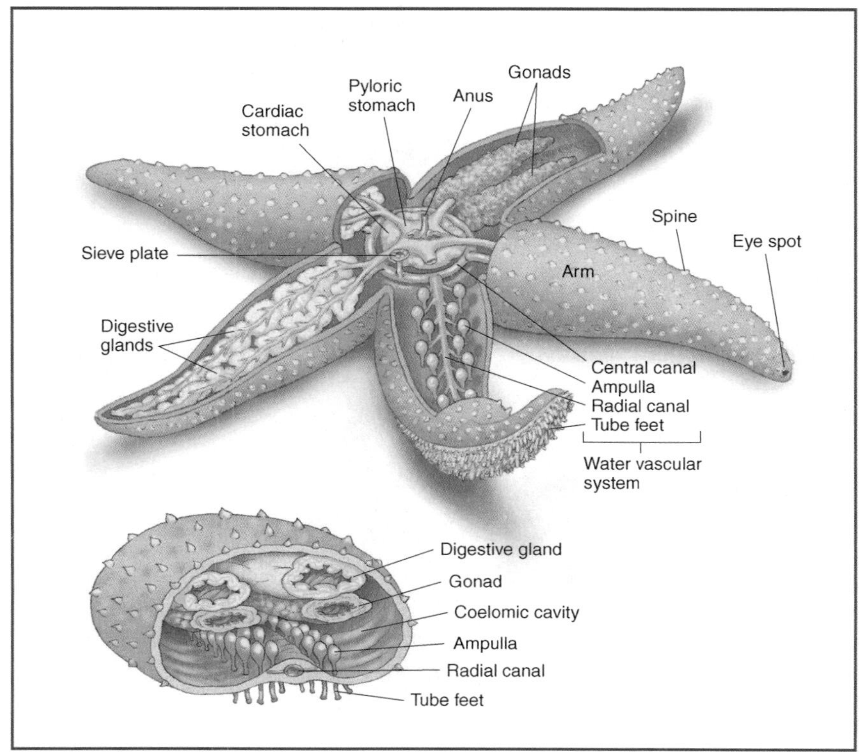

FIGURE 19.6

PHYLUM CHORDATA

This phylum contains three subphyla: Urochordata, Cephalochordata, and Vertebrata. Despite the diversity within this group all chordates have the following characteristics in common, at least during some point in their development: a **notochord**, a **dorsal tubular nerve chord**, pharyngeal pouches, and a **postanal tail.** Figure 19.7 shows the evolutionary relationships among the chordates and some of the evolutionary trends.

Subphylum Urochordata

This group contains the sea squirts or tunicates. The adult of this group lacks the nerve chord and the notochord. Obtain a tunicate larvae slide and examine this with the aid of a microscope. Use the space below to draw what you see.

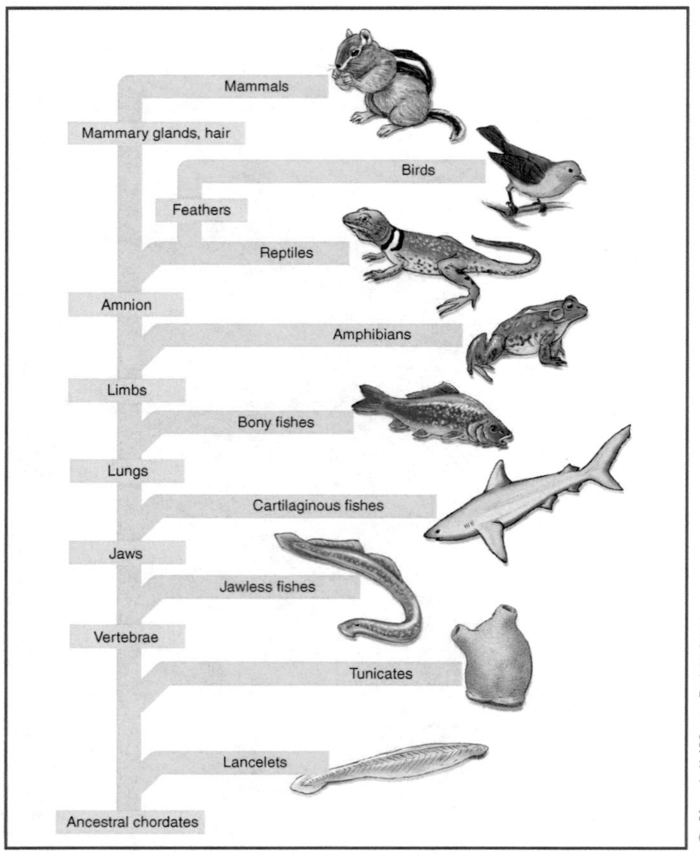

FIGURE 19.7

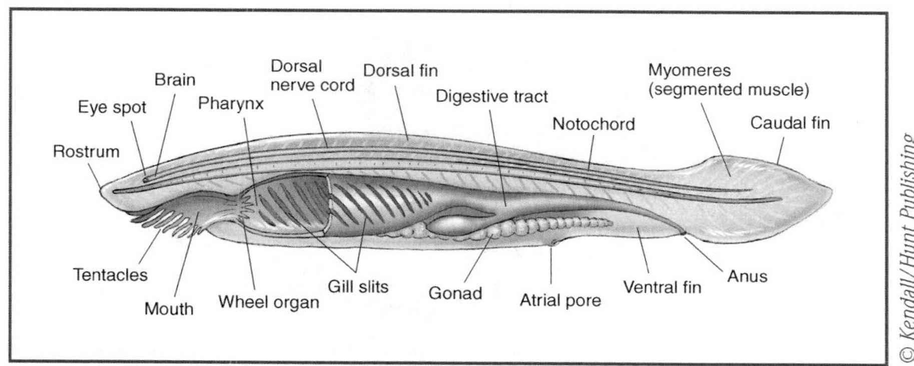

FIGURE 19.8

Subphylum Cephalochordata

This group contains all the typical chordate characteristics as an adult. This phylum contains the lancets which are depicted in Figure 19.8. Obtain a prepared slide of the lancet and observe with the aid of a microscope. Locate and identify all structures shown in Figure 19.8. Use the space below to draw what you see.

Subphylum Vertebrata

The defining characteristic of this group is the replacement of the notochord with a **vertebral column**. They also have well-developed heads, a strong and jointed endoskeleton, a closed circulatory system, and efficient respiration and excretion. This is a diverse group that contains six classes: Chondrichthyes, Osteichthyes, Amphibia, Reptilia, Aves, and Mammalia. The jawless fishes are grouped into a superclass: Agnatha.

Superclass Agnatha

These are the jawless fishes, such as the hagfish and lamprey (Figure 19.9). All other classes of chordates have jaws whose evolution Figure 19.7 shows. Observe any jawless fishes made available by your lab instructor.

Class Chondrichthyes

This group contains the **cartilaginous** fishes, such as sharks (Figure 19.10), skates, and rays. The defining characteristic of this group is the skeleton made of **cartilage**. The best known members of this group are the sharks, much maligned in popular

FIGURE 19.9

FIGURE 19.10

literature and movies. This is clearly a group well-adapted for a predatory lifestyle. Figure 19.11 shows the shark's basic body plan and internal anatomy. Notice the **lateral line system.** This allows sharks to detect prey thrashing in the water. This is why you should never splash violently around sharks. Compare this figure to any preserved specimens provided by your instructor.

Class Actinopterygii

These are the ray-finned fishes (Figure 19.12), which include most fish species that people are familiar with. As the name indicates, this group has paired fins supported by bony rays. They also contain a **swim bladder**, which allows the fish to maintain its position in the water column without actually having to move. Observe any preserved specimens provided by your instructor.

Class Sarcopterygii

These are the lobe-finned fishes, which are characterized by paired lobed fins and also contain a swim bladder. This group is mostly extinct and includes the coelacanth, which we once believed was extinct. This class is the ancestor of the amphibians.

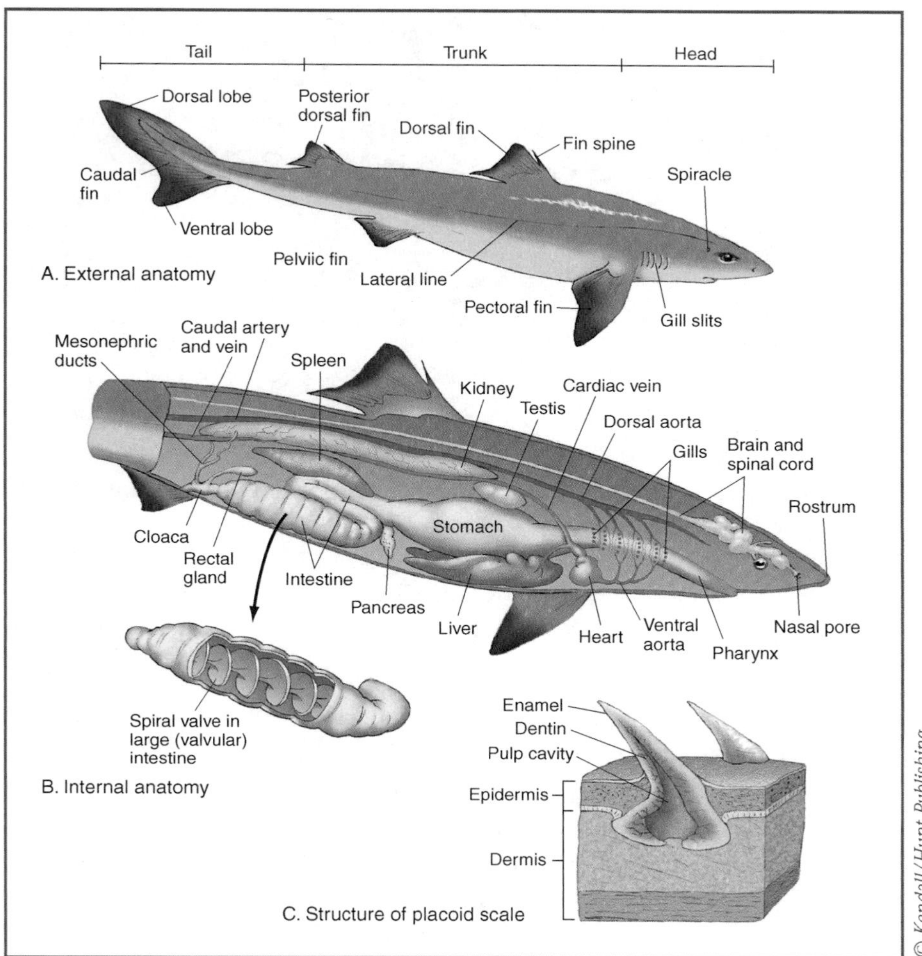

FIGURE 19.11

FIGURE 19.12

Class Amphibia

This group contains the amphibians, which include frogs (Figure 19.13), toads, newts, salamanders, and caecilians. This group contains adaptations for a more terrestrial lifestyle. These include **limbs** for walking, a **tympanum** for hearing, **eyelids, lungs** (in some species), and a **pharynx** used to produce sound, such as the croaking of a frog. All of these adaptations evolved for movement onto land. However, amphibians are not totally independent of an aquatic existence as their larvae are typically aquatic.

FIGURE 19.13

Examine figures 19.14, 19.15, 19.16, and 19.17. These figures show the internal anatomy of a representative amphibian: the frog. Obtain a preserved specimen, a dissection tray, scalpel and/or scissors, and surgical gloves. Cut through the skin of the frog to expose the organs beneath and compare your specimen to the figures. Do not cut too deep as this may damage the structures beneath. The color-coded blood vessels in Figure 19.17 are for illustrative purposes only, and your specimen will not be that color unless it has been injected with latex. Locate and identify structures indicated in Figures 19.14–19.17. You will need to exchange frogs with another lab group to observe the different sex organs.

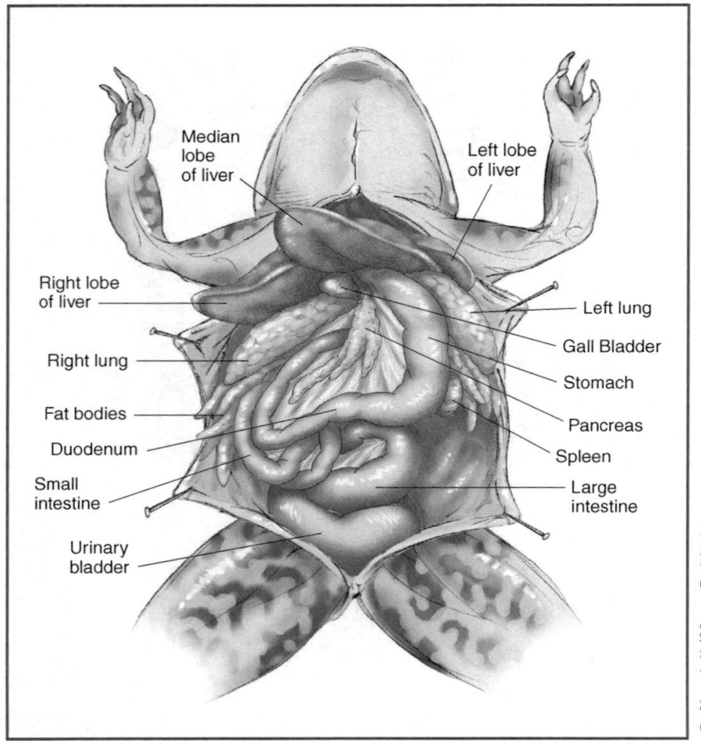

FIGURE 19.14

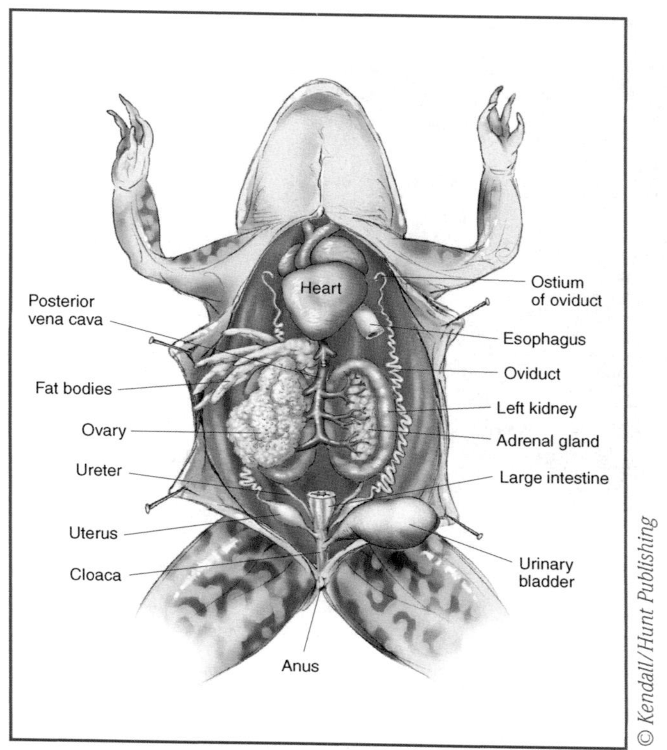

FIGURE 19.15

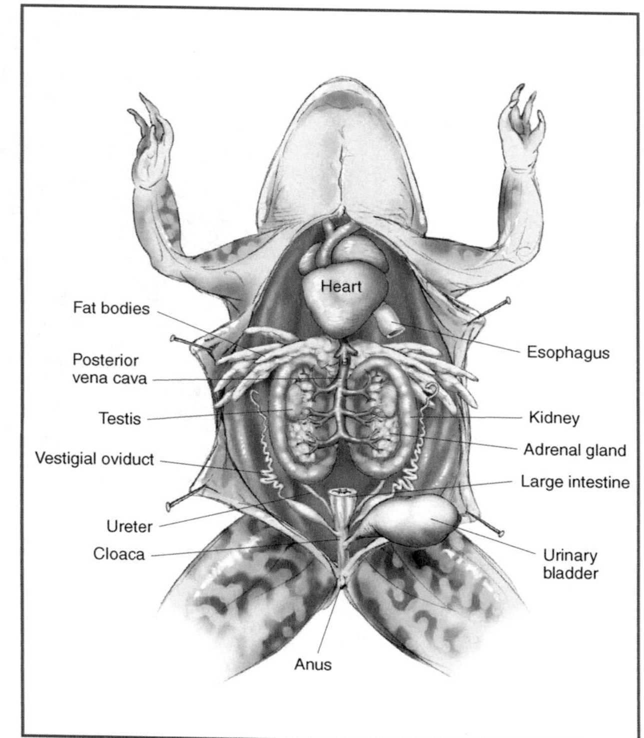

FIGURE 19.16

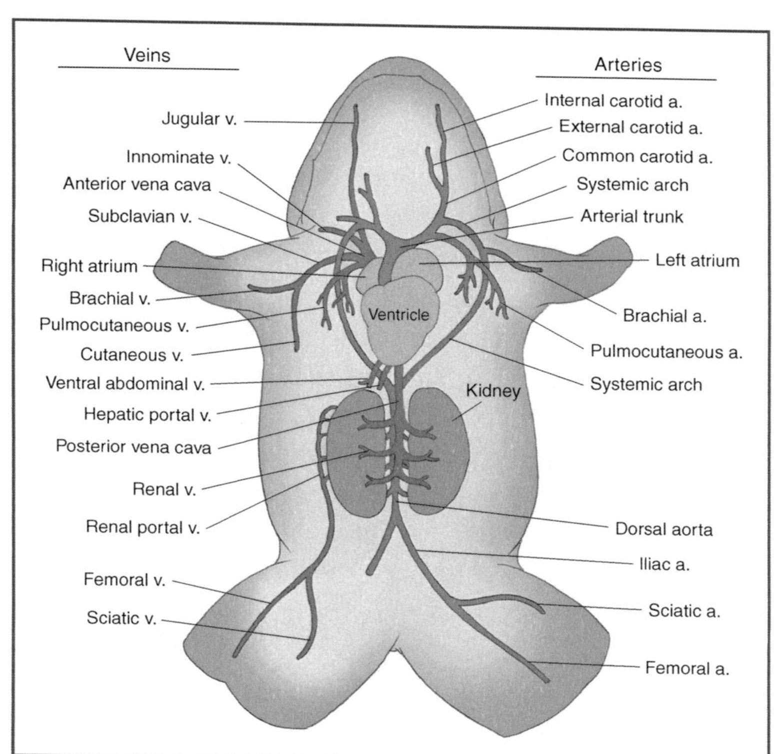

FIGURE 19.17

FIGURE 19.18

Class Reptilia

The reptiles are a diverse group that contains the snakes, crocodilians (Figure 19.18), turtles, and lizards. The characteristics of this group are further modifications for a terrestrial existence. For instance, **scales** cover them that prevent water loss. The most significant adaptation though is their **amniotic egg.** This adaptation has allowed reptiles freedom from an aquatic existence for reproduction. They house the developing embryo in an environment suitable for its protection and development, an adaptation that amphibians lack. Examine any preserved specimens provided by your lab instructor.

Class Aves

Birds compose this group (Figure 19.19). Typical characteristics of birds include **feathers**, a **hard-shelled amniotic egg**, **wings**, a **four-chambered heart**, and **air sacs** for respiration. They are also the first **endothermic** group. This means that they are capable of maintaining a constant internal body temperature via their own metabolism. This allows birds to exist in environments such as Antarctica that would kill reptiles and amphibians such as Antarctica. All other animals (except mammals) are **ectothermic**, which means that the external environment controls their internal body temperature to a large degree. In the past, people have used the terms warm blooded and cold blooded to refer to endothermic and ectothermic organisms. These terms have little biological meaning though as an alligator's internal body temperature may be higher than yours at any given time depending on the ambient temperature. So eliminate warm blooded and cold blooded from your vocabulary at this time.

FIGURE 19.19

This is also the first class to develop a four-chambered heart, which allows for a more efficient circulatory system. The fishes have a two-chambered heart, and reptiles and amphibians have a three-chambered heart: two atria and one ventricle. The amount of division of the ventricles in reptiles varies (as crocodiles also have a four-chambered heart), but you can clearly see the evolution of a four-chambered heart as we progress through the animal kingdom. Observe any preserved specimens your lab instructor provides.

Class Mammalia

Mammals comprise this group, which we belong to. Mammals have **hair**, a **four-chambered heart**, a **diaphragm**, and **differentiated teeth**. They are **endothermic** and have **mammary glands**, which give this class its name. Most mammals are **placental** mammals (Figure 19.20), which provide nourishment to the developing young, gas exchange and elimination of wastes via an **umbilical cord** that connects the embryo to the **placenta**. However, some (the **monotremes**), such as the duck-billed platypus, lay eggs (Figure 19.21), which clearly show the reptilian origins of mammals. The **marsupials** (Figure 19.22) have a pouch called a **marsupium**. In these species the young are born very early and are not fully developed. The young enter the pouch and attach to a nipple to continue development. Observe any preserved specimens provided by your instructor. A future lab exercise will cover the internal anatomy of mammals.

FIGURE 19.20

FIGURE 19.21

FIGURE 19.22

NAME ____

EXERCISE 19

QUESTIONS

1. What are some of the adaptations animals evolved for a terrestrial existence?

2. What are the characteristics of birds?

3. What are the characteristics of mammals?

4. When do we first see the separation of the ventricles beginning to form?

5. In what group of mammals do we see reptilian characteristics such as an egg?

6. What is the most important adaptation of reptiles for a terrestrial existence?

7. At what point do we see the evolution of jaws?

EXERCISE 20

MAMMALIAN ANATOMY

INTRODUCTION

All organisms must maintain some degree of **homeostasis** in order to remain alive. Mammals are no exception, and as a result they have highly organized bodies to efficiently maintain a relatively constant internal environment. They accomplish this by organizing cells into **tissues**, tissues into **organs**, and organs into **organ systems**. Today you will learn these organs by examining mammalian anatomy through the dissection of a fetal pig.

EXTERNAL ANATOMY

Obtain a tray, scalpel and/or scissors, a dissecting tray, and surgical gloves from your instructor. Put on your gloves and obtain a fetal pig. Examine Figure 20.1. This figure shows the external anatomy of a fetal pig. Pigs are mammals and as a result have certain characteristics shared by all mammals. These characteristics include both **hair** and **mammary glands**. Mammals are also **placental** mammals, which mean that a **placenta** forms during **gestation**. This structure facilitates the development of the fetus by exchanging nutrients, gases, and wastes with the mother via an **umbilical cord** that should be evident in your specimen. You also had an umbilical cord attached at your navel. Mammals, as with many other groups of animals, also have separate sexes. Figure 20.1 shows the external differences between these two sexes.

ORAL CAVITY AND PHARYNX

Open the mouth of the fetal pig as wide as you can and cut its cheeks on both sides as indicated in Figure 20.2. Hold the now cut mouth open as wide as you can and

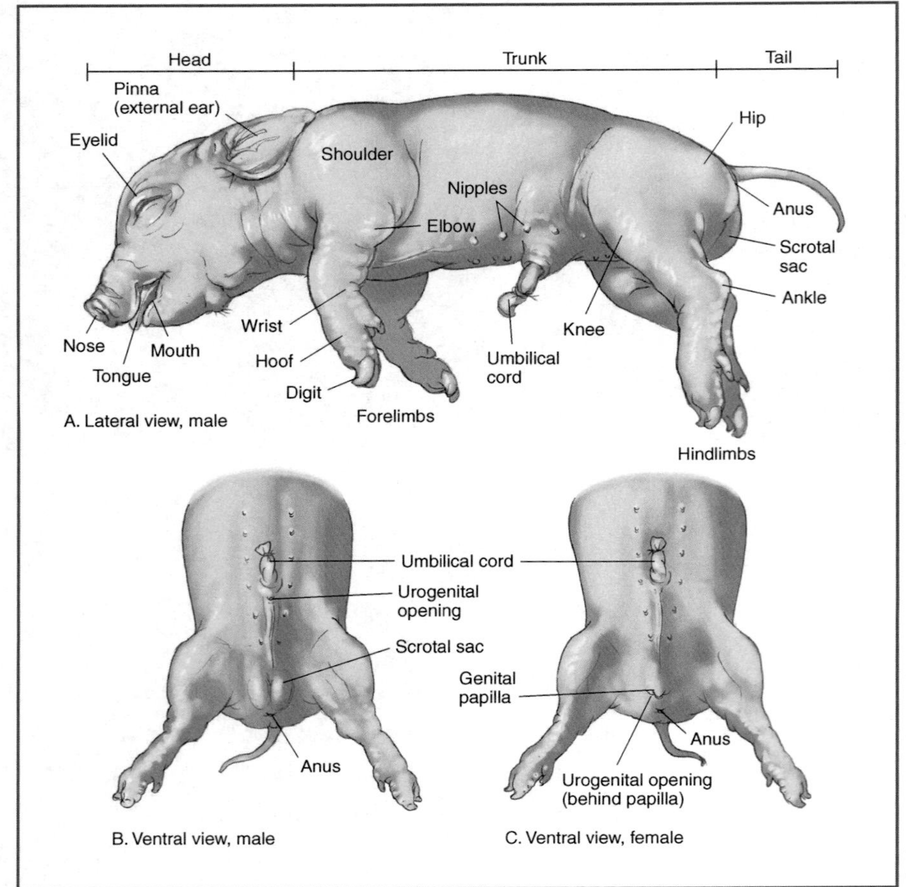

FIGURE 20.1

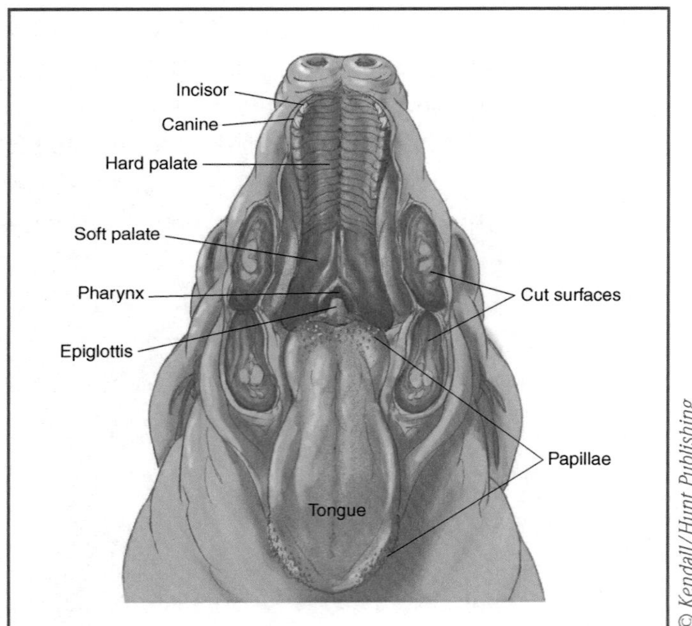

FIGURE 20.2

locate all structures listed in Figure 20.2. If you open the mouth wide enough and insert your fingers to push down the tongue, you should be able to see the **epiglottis**. The epiglottis prevents food from entering the respiratory tract when the pig swallows.

DISSECTION

Examine Figure 20.3. This figure shows the location and order of cuts that you should make in order to fully examine the organs within. You may need to tie down the legs in order to gain access to the ventral side of your specimen. When making your cuts be careful to penetrate just the tissue of the skin. Do not jab your instruments into the pig as this may damage organs just underneath. Fluid may run out at this time. This is the preservative used to maintain the integrity of the specimen and is normal. Once you have made the cuts, you will need to pull back the skin to fully expose the organs beneath.

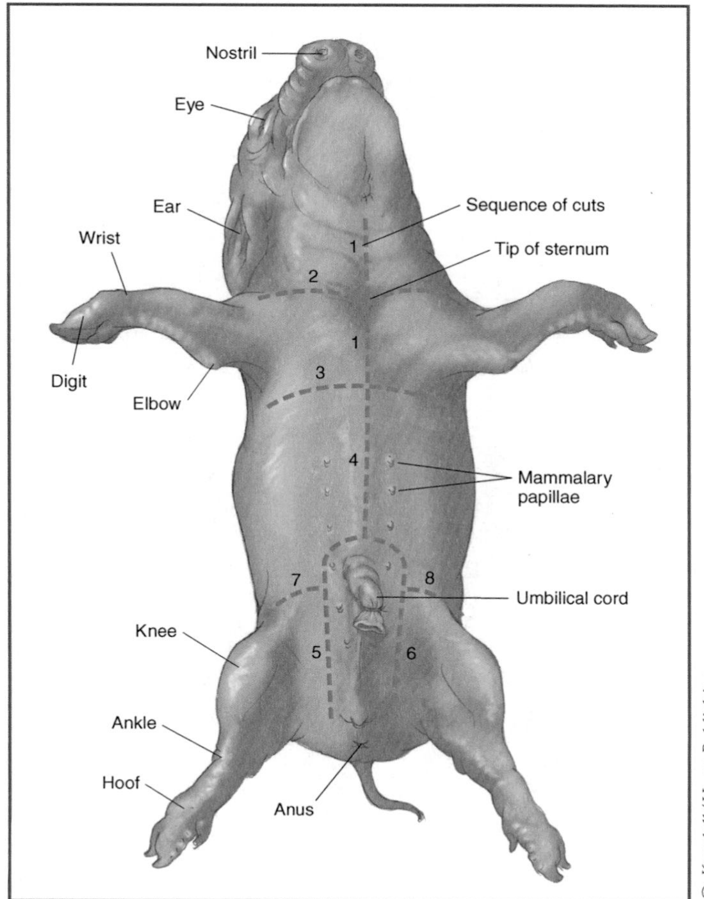

FIGURE 20.3

ORGANS

Examine Figure 20.4. This figure shows the major organs you will see once you have opened the **thoracic** and **abdominal** cavities. Beginning in the neck area you should be able to see the **larynx** followed by the **trachea**. White rings of cartilage circle the trachea, designed to keep the passageway open for breathing. If you pull the trachea back you should be able to see the **esophagus,** which will be collapsed in your specimen. Located on and around the trachea are the **thyroid** and **thymus** glands. It is possible that parts of these glands will pull off as you pull the skin back. Next you will find the **heart** surrounded on both the right and left sides by the **lungs**. Posterior to the lungs you should notice a sheet-like muscle called the **diaphragm** that separates the thoracic and abdominal cavities. It also helps expand the lungs during breathing. The diaphragm is unique to mammals.

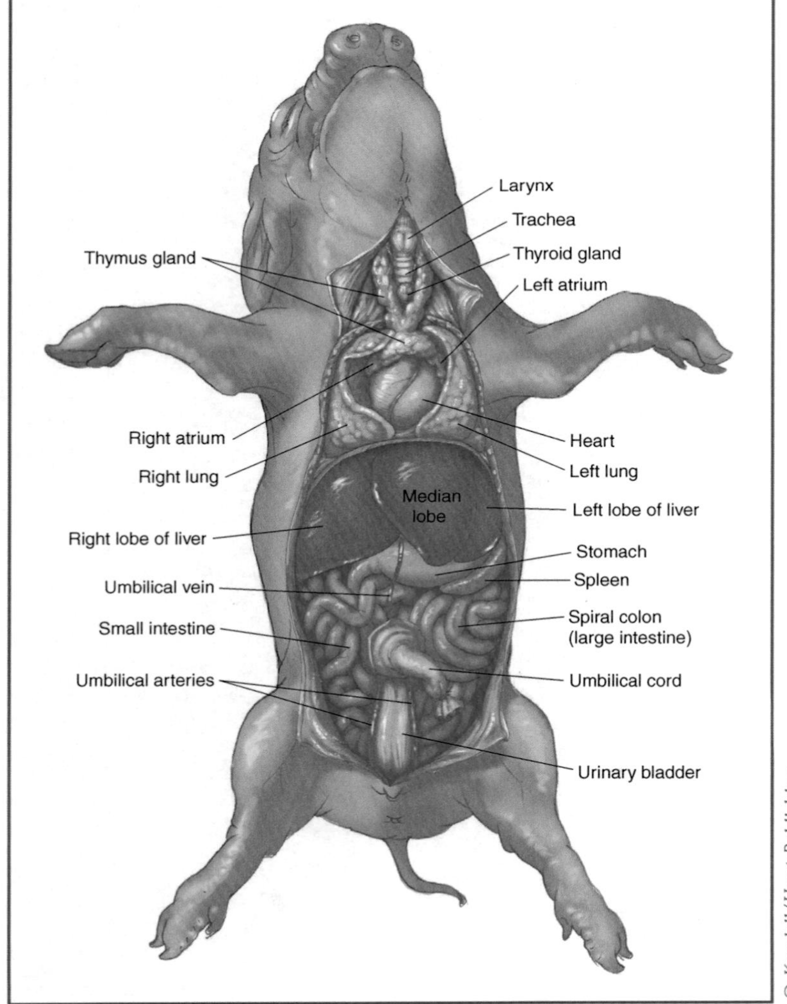

FIGURE 20.4

The first major organ you will notice in the abdominal cavity is the **liver.** This is a multi-lobed organ and is hard to miss. Located posterior to the liver is the **stomach.** The **spleen** lies on the left side of your pig lateral to the stomach. Also, you will notice both the small and large **intestines.** You should also see the **urinary bladder.** The **kidneys** may not be readily visible, but if you pull back the intestines you should locate both kidneys on the dorsal wall of the abdominal cavity. Figure 20.5 shows the location of the kidneys.

Using your textbook as a guide, indicate the functions of the following structures:

Larynx

Trachea

Thyroid gland

Thymus gland

Heart

Lungs

Diaphragm

Liver

Spleen

Small intestine

Large intestine

Kidneys

Bladder

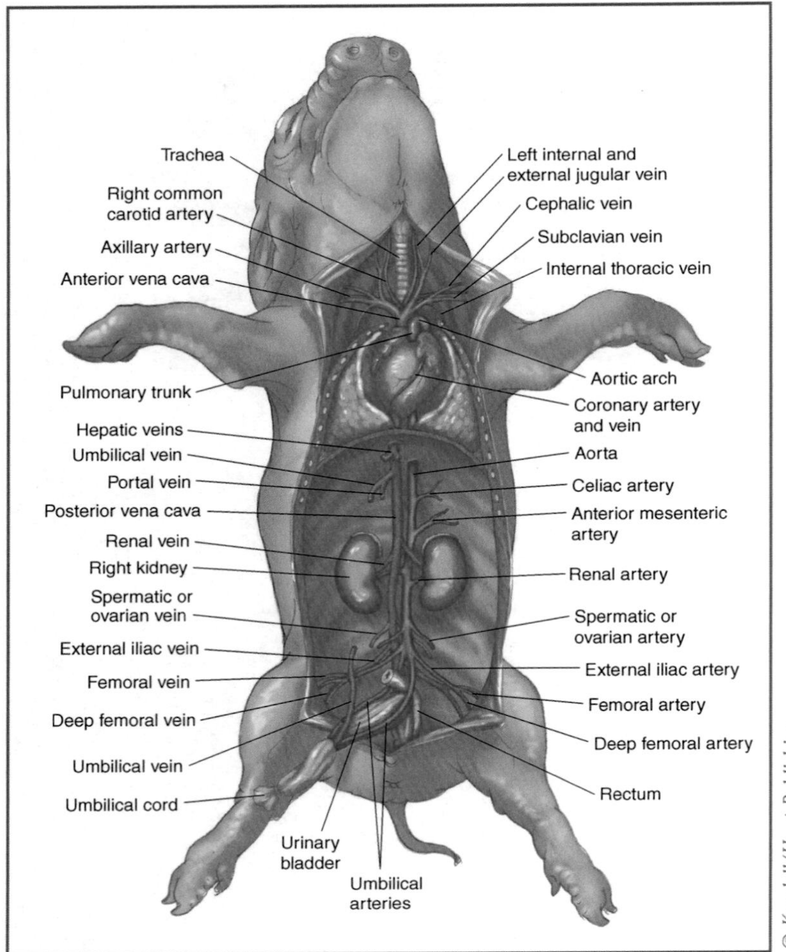

FIGURE 20.5

CIRCULATORY SYSTEM

Examine Figure 20.5. This figure shows the circulatory system. On the diagram, arteries appear in red and veins in blue. In your fetal pig the arteries are red and the veins are blue if they were injected with red and blue latex respectively. Arteries carry oxygen-rich blood from the heart to the tissues of the body, and veins carry oxygen-poor blood away from tissues; hence the two color choices. This is a learning aid only. In a living organism the veins and arteries do not take on different colors.

In order to see the veins and arteries you will need to pull back the organs to view these deeper structures. Look at the heart and notice the artery that is exiting this structure. This is the **aortic arch**, and oxygenated blood leaves the heart through this pathway. The large vein located near the right atrium of the heart is the **vena cava.** The fetal pig has an anterior and posterior vena cava. Oxygen-poor blood enters the heart through this structure. Locate the trachea. Located on both sides of the trachea are arteries and veins. The **carotid artery** runs along both sides of the trachea, and the vein that runs parallel to this is the **jugular vein**. The jugular vein has two branches on each side: the internal and the external jugular vein. Locate both of these branches.

Examine the abdominal cavity. You should notice one large vein and one large artery running the length of the abdominal cavity. The vein is the **posterior vena cava**, and the artery is the **aorta**. You can locate various arteries and veins in the abdominal cavity by locating branches off of these two structures. For instance, you should be able to locate the **celiac artery** branching off of the aorta first followed by the **anterior mesenteric artery**. Entering the kidney is the **renal artery**, and exiting the kidney is the **renal vein**. The renal artery sends blood to the kidney from the aorta, and the renal vein sends blood to the posterior vena cava.

Branches of both the aorta and the posterior vena cava enter the hind legs. These branches run near the femur of the leg, and as a result are named accordingly the **femoral artery** and **femoral vein**. There are also deep femoral arteries and deep femoral veins, which are branches off the femoral artery and femoral vein respectively.

Examine the umbilical cord. This structure should contain both an **umbilical artery** and an **umbilical vein**. These two structures connect to the placenta via the umbilical cord.

UROGENITAL SYSTEM

The urogenital system is actually two systems combined: the **urinary system** and the **reproductive system**. Depending on the sex of your specimen some structures are going to differ. Examine Figures 20.6 and 20.7. These show the urogenital system of both females and males respectively. Examine the urogenital system of your pig. This may require some additional dissection if you did not originally cut down far enough to examine these structures. Once completed, switch pigs with another group that has a fetal pig of the opposite sex.

After examining the urogenital systems of the sexes, list what they have in common and what the differences are.

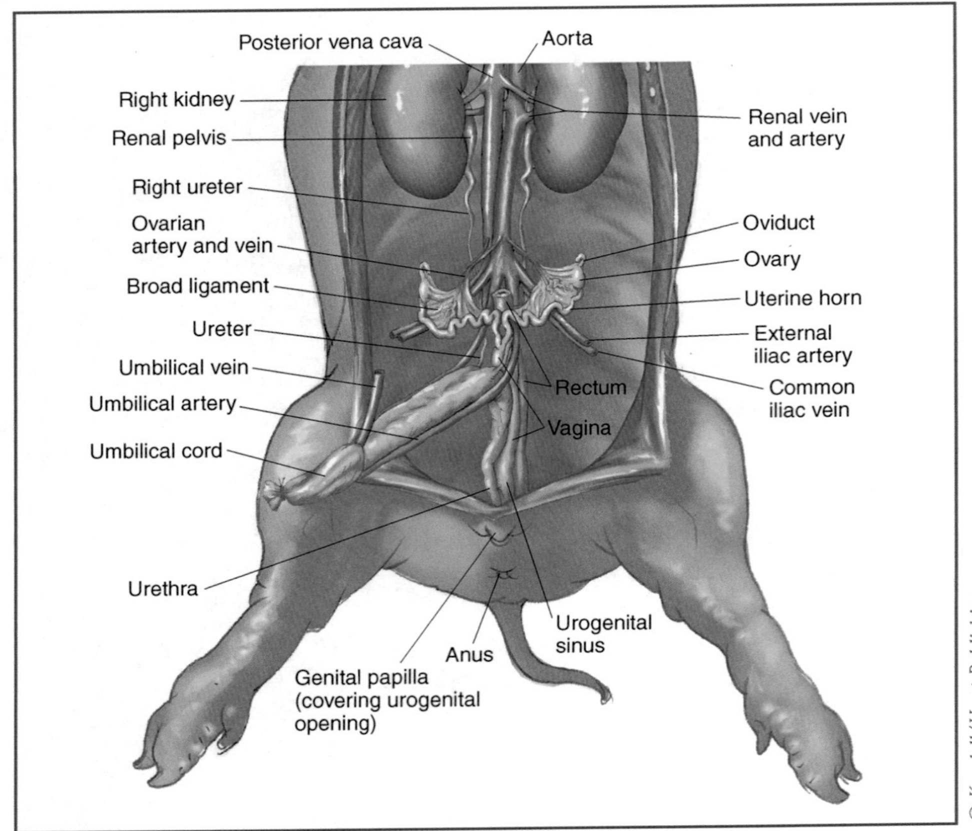

FIGURE 20.6

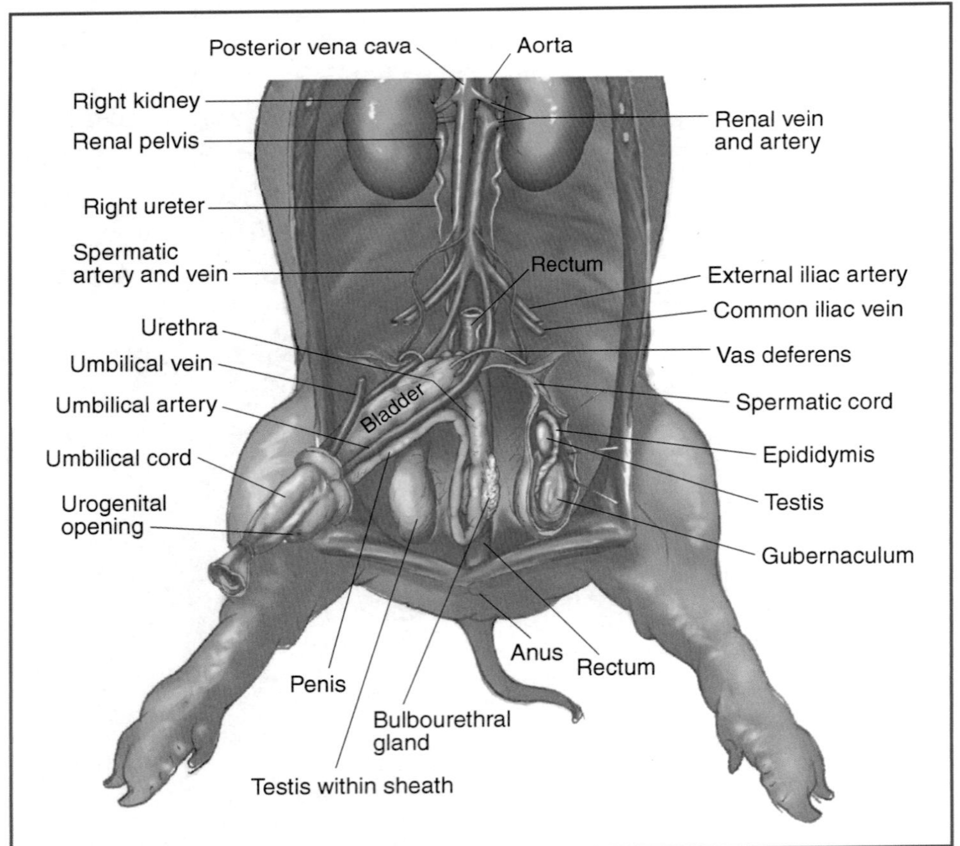

FIGURE 20.7

NAME ____________________

EXERCISE 20

QUESTIONS

1. What is the name of the artery that enters the kidney?

2. What is the first branch off of the aorta as it enters the abdominal cavity?

3. The urogenital system is made of what two systems combined?

4. What does it mean to be a placental mammal?

5. What are the functions of the following? Use your textbook if needed.

Kidneys

Spleen

Trachea

Larynx

Liver

Gall bladder

EXERCISE 21

CARDIOVASCULAR SYSTEM

INTRODUCTION

The cardiovascular system is one of several systems in the body designed to maintain homeostasis. This system in vertebrates can be quite complex with **arteries, arterioles, veins, venules, capillaries**, and the **heart**. All of these structures are designed to either transport blood or pump blood to the various tissues within the body. We will begin our coverage of the cardiovascular system with the heart.

THE HEART

The heart is the pump of the circulatory system, and Figures 21.1 and 21.2 show its external anatomy. The essential purpose of the heart is to receive oxygen-poor blood from the body, to transport oxygen-poor blood to the lungs, to receive oxygen-rich blood from the lungs, and to send oxygen-rich blood to the tissues. The heart accomplishes this task through a series of chambers and valves shown in Figure 21.3.

The hearts of mammals and birds have four chambers, which are the right and left **atria** and the right and left **ventricles**. The atria receive blood from either the **vena cava** in the case of the right atrium or the **pulmonary veins** in the case of the left atrium. Both atria pump blood into the ventricles. The right ventricle sends oxygen-poor blood to the lungs via the **pulmonary arteries**, and the left ventricle pumps blood to the body via the **aorta**.

Valves separate the atria from the ventricles. Valves also separate the ventricles from the pulmonary artery and veins and prevent the backflow of blood when the atria and ventricles relax. For instance, the **tricuspid valve** separates the right atrium from the right ventricle. When the atrium contracts, the valve opens and

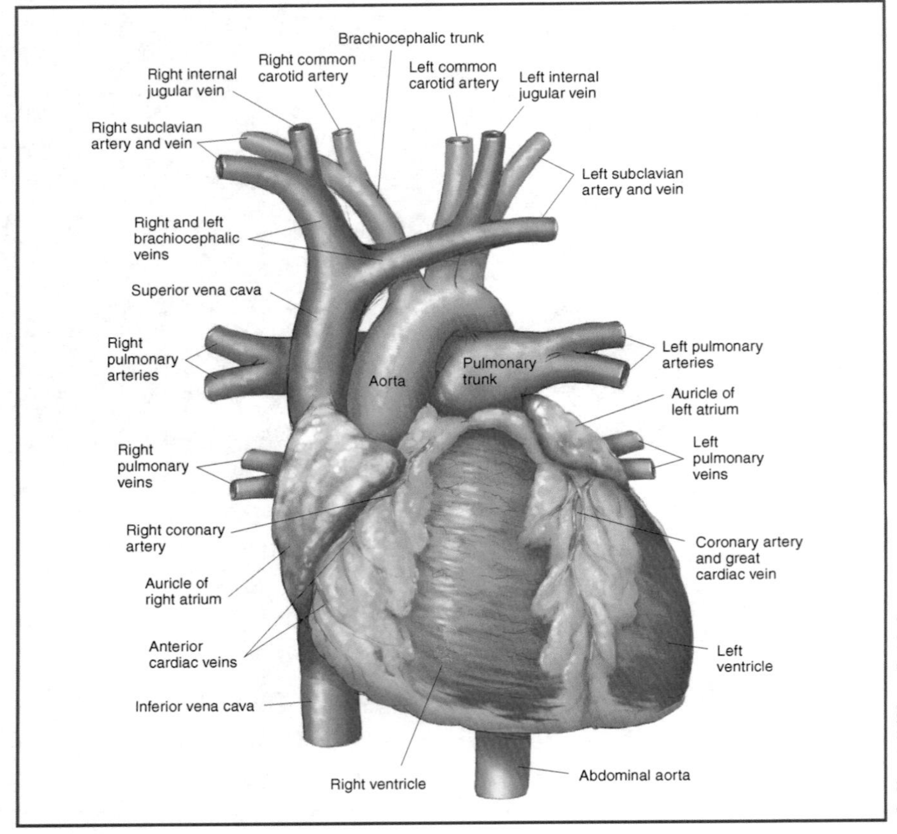

FIGURE 21.1

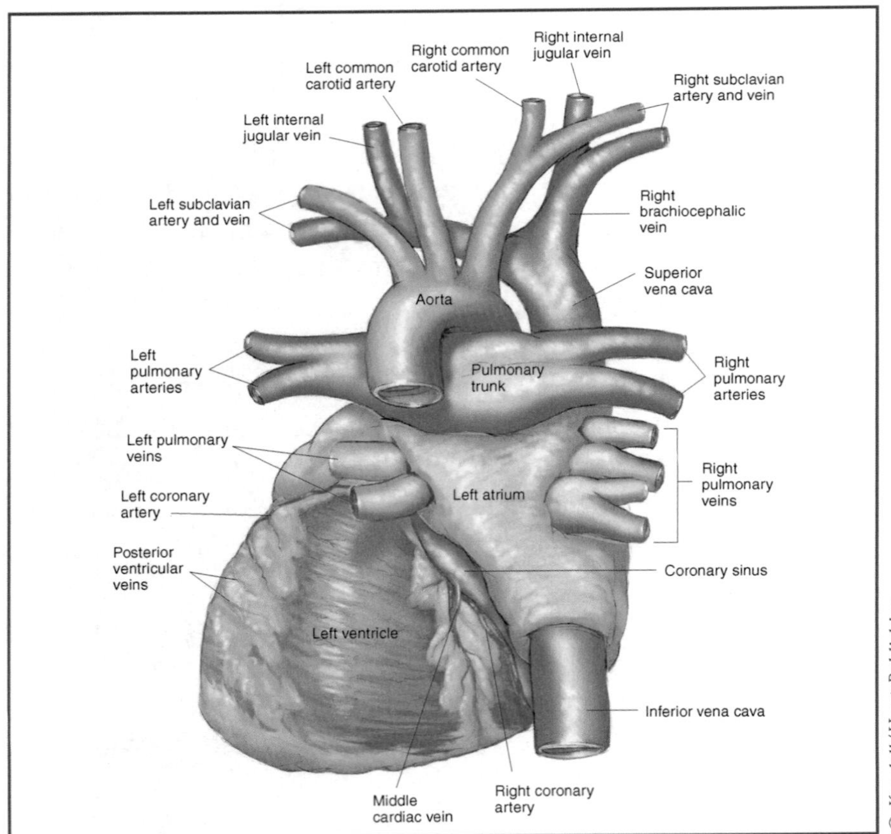

FIGURE 21.2

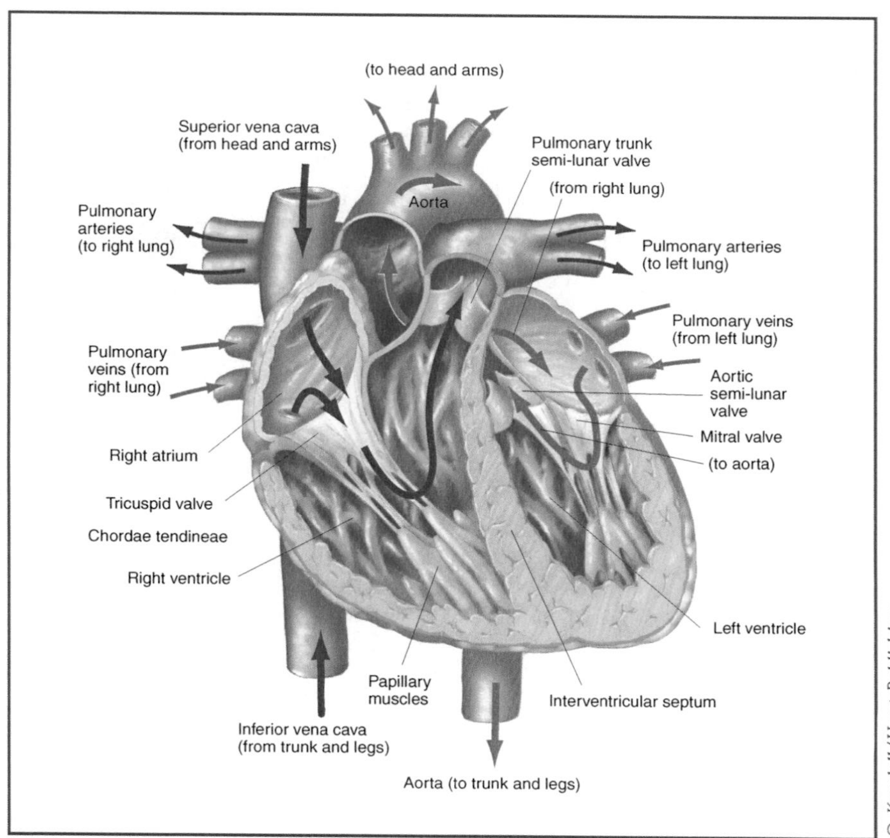

FIGURE 21.3

blood enters the right ventricle. However, when the atrium relaxes and the ventricle contracts, the valve closes to prevent blood from entering the atria from the ventricle. Instead, the **pulmonary semi-lunar valve** opens, and blood flows from the right ventricle into the pulmonary artery on the way to the lungs. The **bicuspid valve** separates the left atrium from the left ventricle, and the **aortic semi-lunar valve** separates the right ventricle from the aorta. These valves work the same way the tricuspid valve and the pulmonary semi-lunar valve operate. Examine Figure 21.3 and notice the tricuspid and bicuspid valves. Attached to these valves are chords called **chordae tendoneae**, attached to **papillary muscles.** These structures prevent the valves from popping open in the opposite direction when the ventricles contract.

Obtain a dissection tray, scissors, gloves, and a heart to dissect. Locate the external structures indicated in Figures 21.1 and 21.2. To examine the internal structures indicated in Figure 21.3 begin cutting at the apex of the ventricle and cut upwards toward the atria. Do not cut the heart into two separate parts. Locate and identify all structures listed in Figure 21.3. If you take a blunt probe you can pass it through the vena cava and aorta to see the openings into the right atria and left ventricle respectively.

CIRCULATION

Figure 21.4 shows the path of blood flow through the heart. Use the space below to trace the path of blood through the heart. Begin with the vena cava, end at the aorta, and include the path to the lungs and back. Be sure to include all valves that the blood passes through.

Arteries and veins are the pathway for blood to travel throughout the body. **Arteries** carry oxygen-rich blood from the heart to the tissues, and veins carry oxygen-poor blood from the tissues to the heart. Arteries branch into **arterioles**, which in turn are connected to **venules** through structures called **capillaries**. This is where the exchange of gases between blood and tissues actually occurs. **Venules** transport

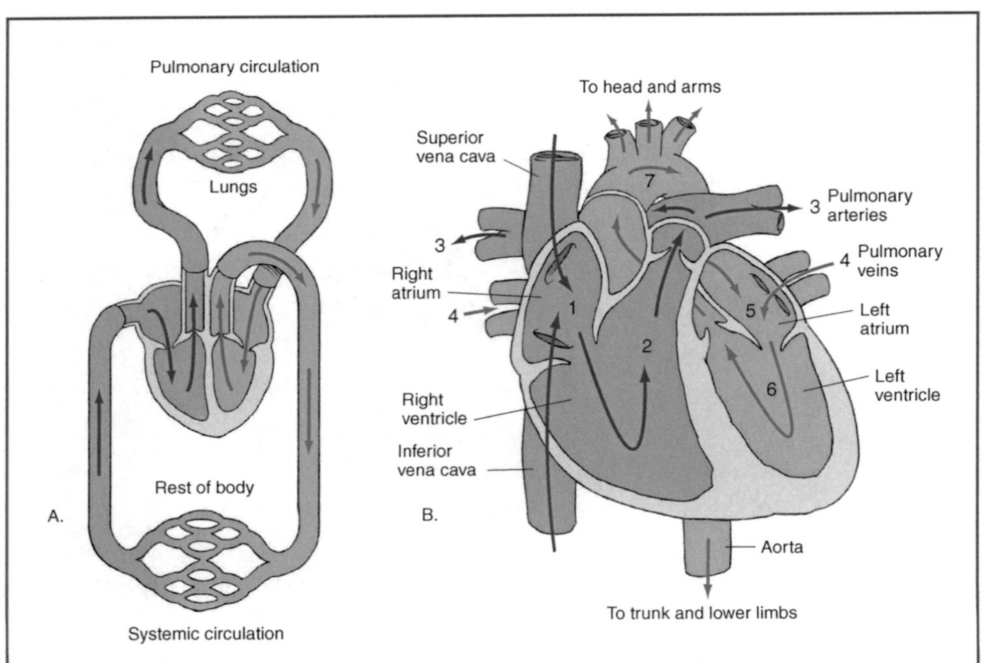

FIGURE 21.4

blood to **veins**. Typically we depict arteries and veins as red and blue respectively; however, the pulmonary artery is typically blue because it carries oxygen-poor blood, and the pulmonary veins are red because they carry oxygen-rich blood. This coloration is for illustrative purposes only for arteries and veins do not take on different colors in organisms. Figure 21.5 shows the major arteries and veins in humans.

Answer the following questions:

1. If you examine the figure of the heart you will notice three branches coming off of the aorta. Examine Figure 21.5 and indicate what arteries those branches lead to.

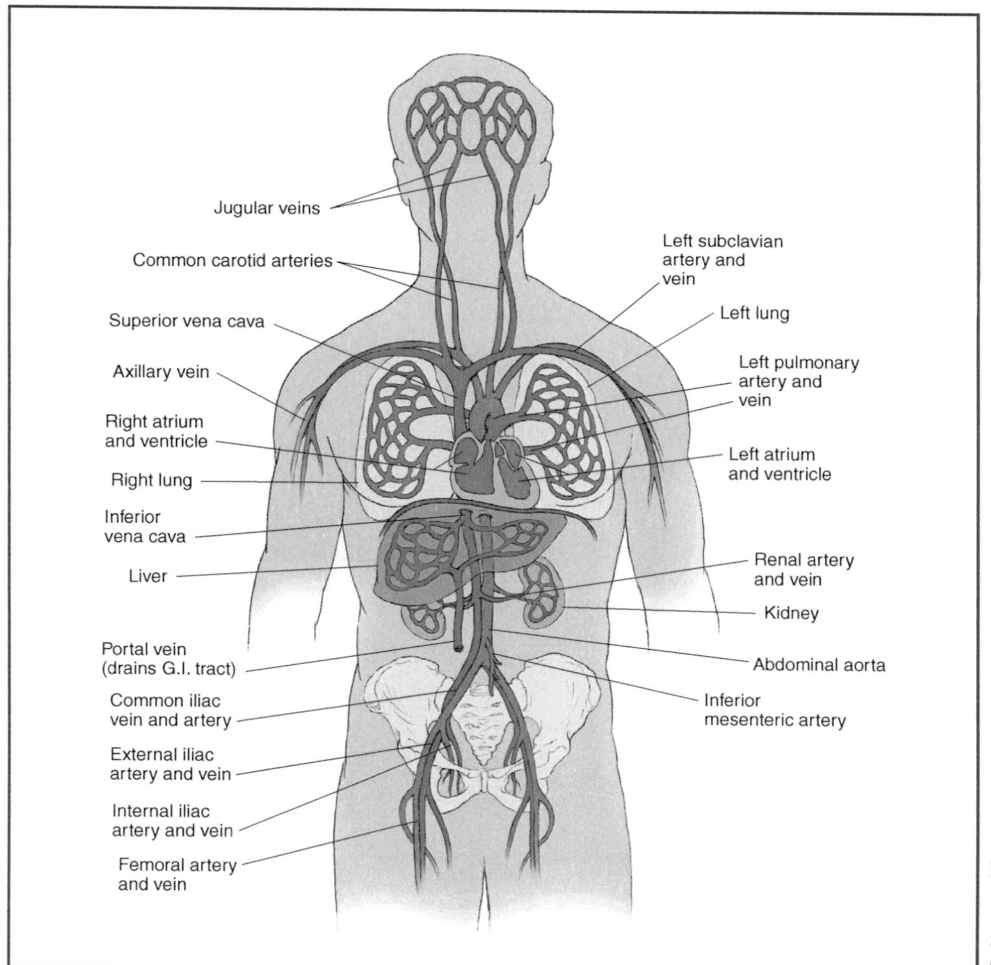

FIGURE 21.5

2. As stated earlier, birds and mammals have a four-chambered heart; however, reptiles (except crocodilians) and amphibians have a three-chambered heat (two atria and one ventricle). This allows oxygen-rich and oxygen-poor blood to mix. Fish species have a two-chambered heart (one atria and one ventricle). What advantages do birds and mammals have from a four-chambered heart compared to reptiles, amphibians, and fish?

BLOOD

The formed elements of blood include red and white blood cells and platelets. Red blood cells consist of **hemoglobin**, which is a molecule that binds to oxygen at the lungs and releases it at the tissues via diffusion. Red blood cells do not have a nucleus at maturity. They look a little bit like a doughnut underneath the microscope, although the "hole" in the middle is really just a depression where the nucleus should be. White blood cells are not really white, but since they do not contain hemoglobin they are not red. They also have a nucleus that can appear quite large with respect to the cell itself. There are a variety of white blood cell types and are used in immune responses. Platelets are used for **clotting. Plasma** is the liquid portion of blood, and it has a straw color. Obtain a prepared blood slide and examine this with the aid of a microscope. Be sure and locate white blood cells, which are less numerous than their red counterparts.

NAME ______________________________

EXERCISE 21

QUESTIONS

1. Draw a heart and label all valves and chambers.

2. What is the difference between pulmonary arteries and veins?

3. Which two animal groups contain 4-chambered hearts?

4. Define hemoglobin.

EXERCISE 22

NERVOUS SYSTEM AND SENSES

INTRODUCTION

One of the characteristics of life is the ability to respond to a stimulus. The nervous system receives input from our senses and integrates that information to produce the appropriate response. Obviously the response you produce to someone giving you a gift will differ from one you give to a lion attacking you. However, it does not matter whether the stimulus is pleasant or unpleasant as the basic pattern is the same: receive information, integrate that information, and produce a response. The only major difference is the response itself. Today we will explore part of the nervous system by examining the brain and some of your senses responsible for detecting environmental signals: hearing and sight.

THE BRAIN

The central control unit for your nervous system is the **brain** (as well as the spinal cord). This structure is responsible for your personality, behavior, intelligence, etc., in other words, the way you view the world and interact with it. The structure of the brain varies across the animal kingdom, but even distantly related animals have similar structures that indicate common ancestry at some point in evolutionary history. Figure 22.1 shows the evolution of the vertebrate brain. While the shape of the structures is different among these groups, you will notice they contain many of the same features. This diagram shows one view only so all similarities are not visible. For instance, while cranial nerves do not appear for the large mammal

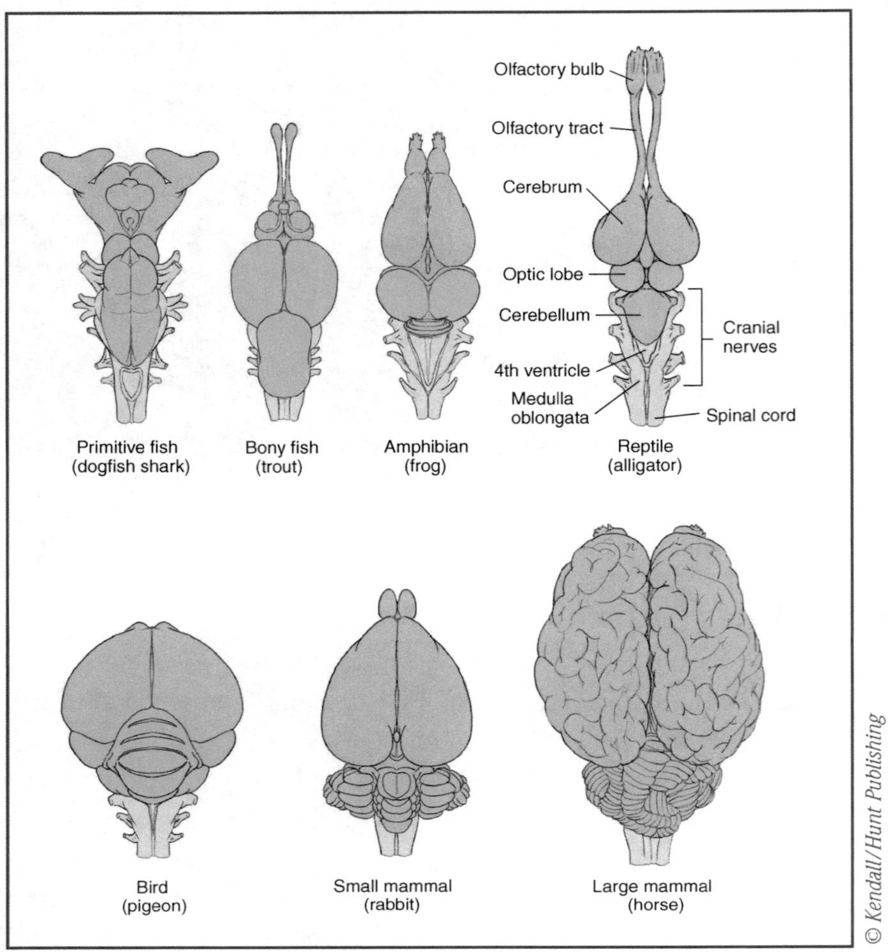

FIGURE 22.1

brain, they are present. Use this figure as a guide to list the structures they have in common in the space below.

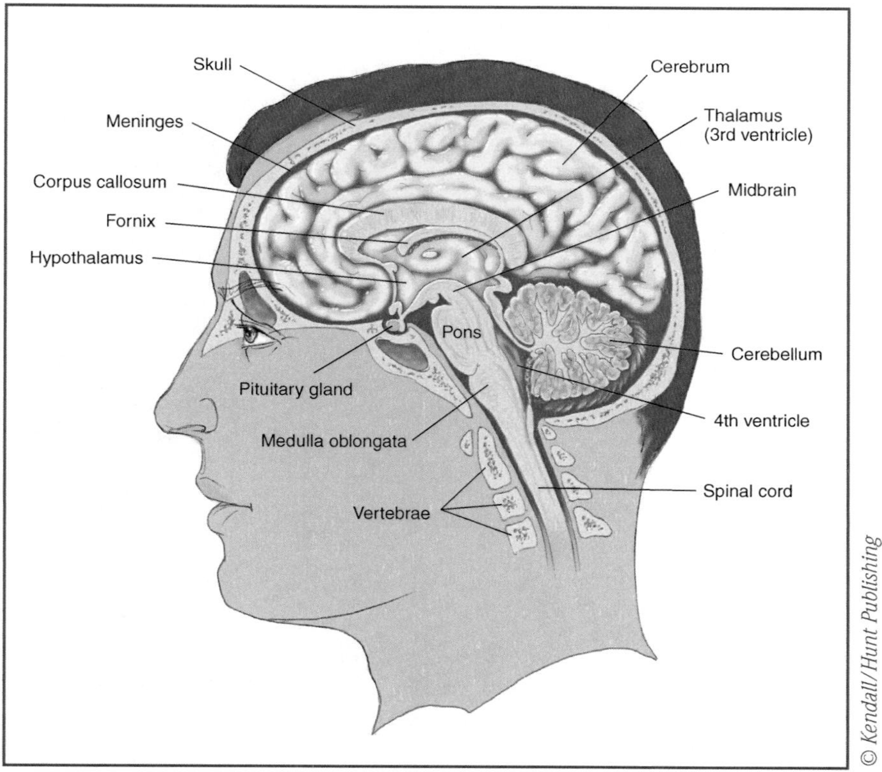

FIGURE 22.2

Figure 22.2 shows a cross section of the human brain. Use models provided to you by your instructor to locate and identify all structures on the model with this figure as a guide. Use your textbook as a reference to define or list the functions of the following structures:

Meninges

Corpus callosum

Hypothalamus

Cerebrum

Thalamus

Midbrain

Cerebellum

Pituitary gland

Medulla oblongata

Pons

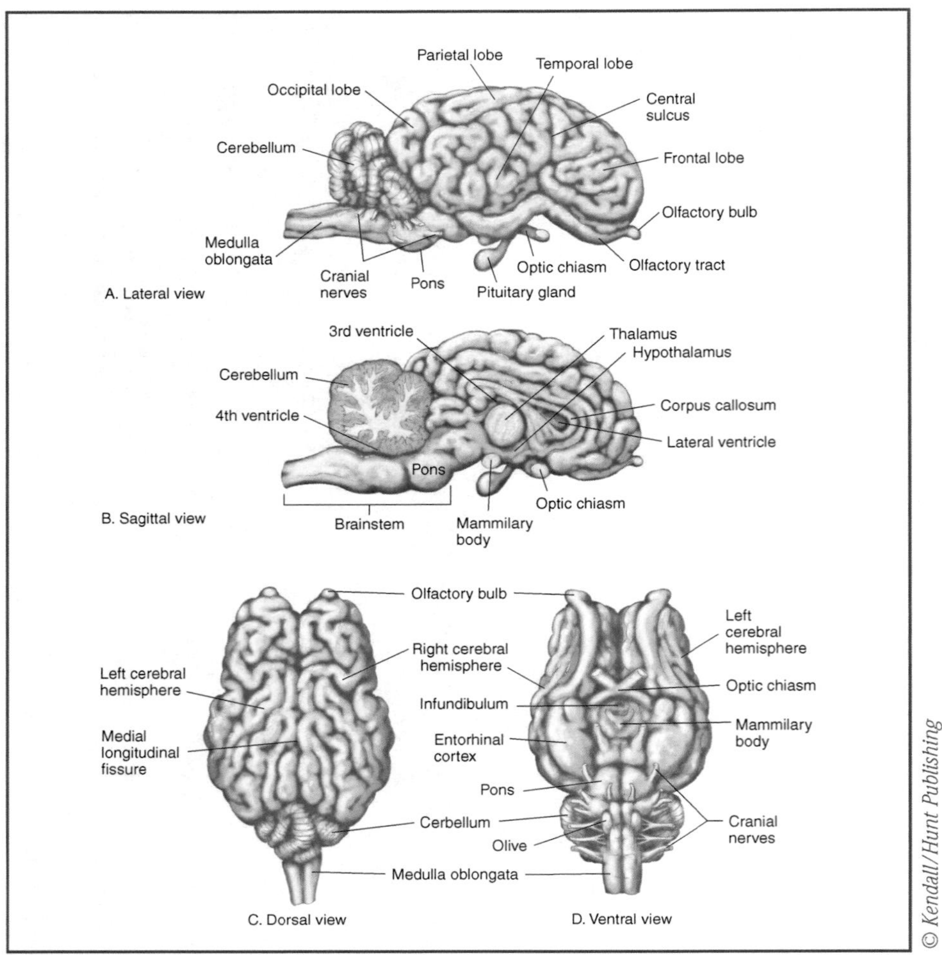

FIGURE 22.3

SHEEP BRAIN DISSECTION

Obtain a dissecting tray, scissors, gloves, and a sheep's brain provided to you by your instructor. Compare the structures of the specimen you possess and compare it to Figure 22.3.

HEARING

Obviously the structures responsible for your ability to hear are your ears. Ears are also largely responsible for your ability to maintain your balance. The parts of your ear and their functions appear below.

Auditory canal—This structure leads to the tympanic membrane.

Cochlea—This has a shape like a snail's shell within the inner ear.

Eustachian tube—This extends from the middle ear to the nasopharynx and equalizes the pressure on the tympanic membrane.

Incus—This is the second of three bones in the middle ear that receives vibrations from the tympanic membrane, sometimes called the anvil.

Malleus—This is the first of three bones in the middle ear that receives vibrations from the tympanic membrane. This structure is sometimes called the hammer.

Pinna—This is the outer part of the ear that channels sound waves to the tympanic membrane.

Round window—An opening between the middle and inner ear.

Semicircular canals—Located within the inner ear and shaped like a tube. Used for rotational equilibrium.

Stapes—This is the third of three bones in the middle ear that receives vibrations from the tympanic membrane. This structure is sometimes called the stirrup.

Tympanic membrane—This is the eardrum; it receives air vibrations.

Label Figure 22.4 using these descriptions as a guide. Use your figure to locate those structures on models provided by your instructor.

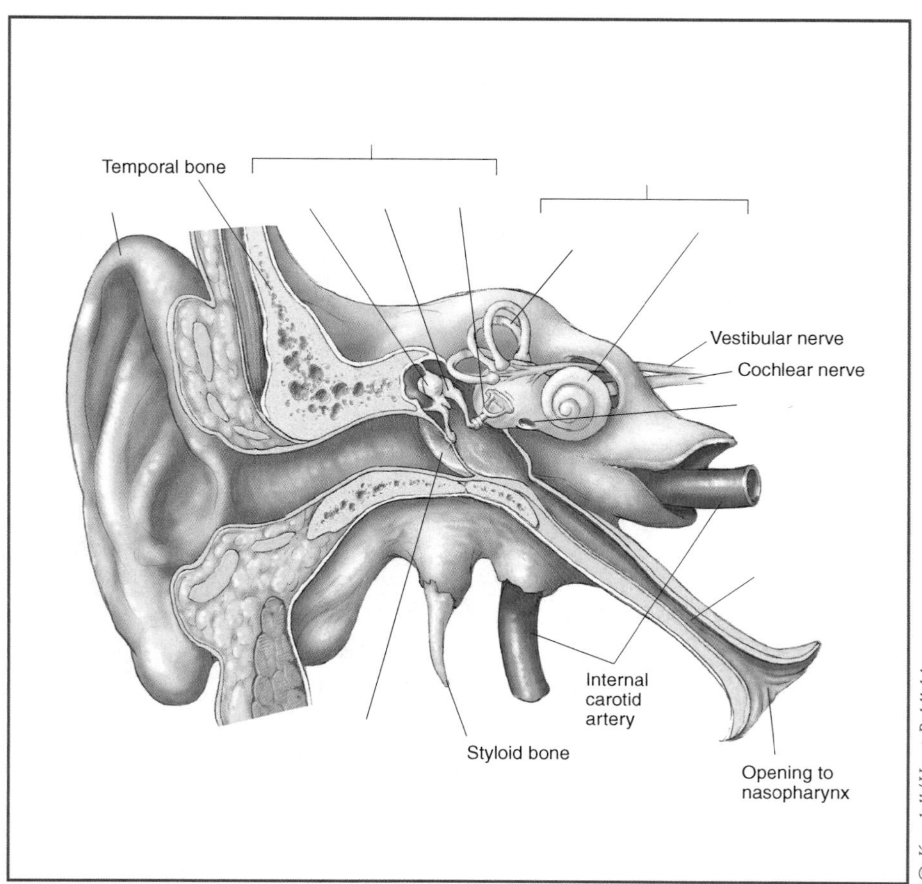

FIGURE 22.4

EYES

As humans, we rely on our eyesight as our principle means of exploring our world. Figure 22.5 gives the basic structure of the eye. Locate the structures listed in this figure on models provided to you by your instructor. Use your book as a reference in order to list the definitions or functions of the following structures:

Pupil

Retina

Iris

Lens

Sclera

Choroid

Fovea centralis

Optic nerve

Vitreous humor

Obtain a dissecting tray, utensils, gloves, and a preserved eye for dissection. Using scissors cut the eye open and examine the internal structure using Figure 22.5 as a guide. Make sure you remove the lens for examination.

Figure 22.6 shows the action of the lens of the eye when focusing. Notice that the lens bends light, and the object projects upside down on the retina. Also notice the shape of the lens when focusing on close and distant objects.

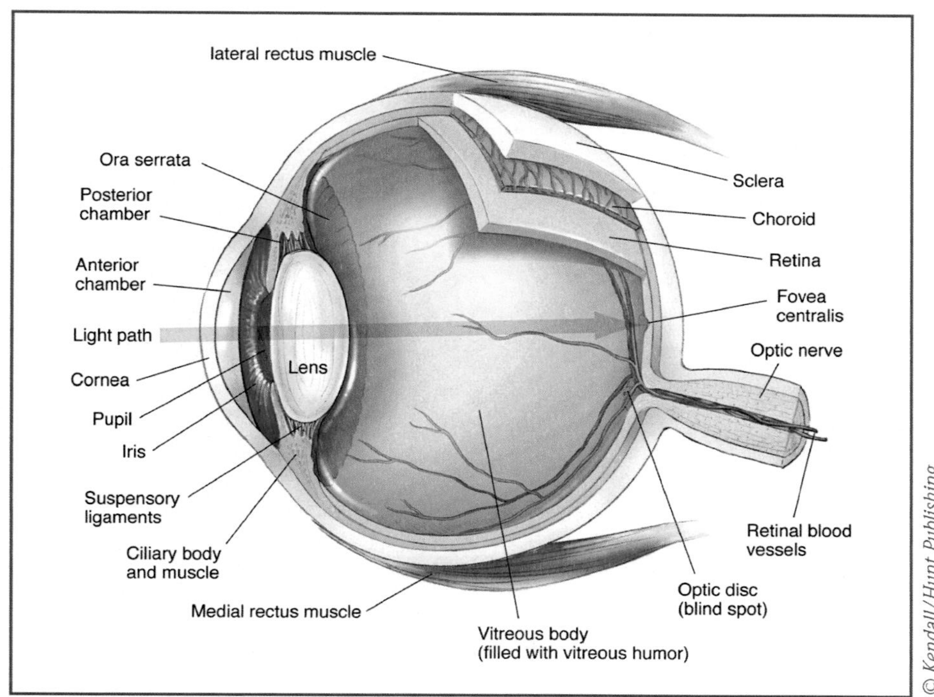

FIGURE 22.5

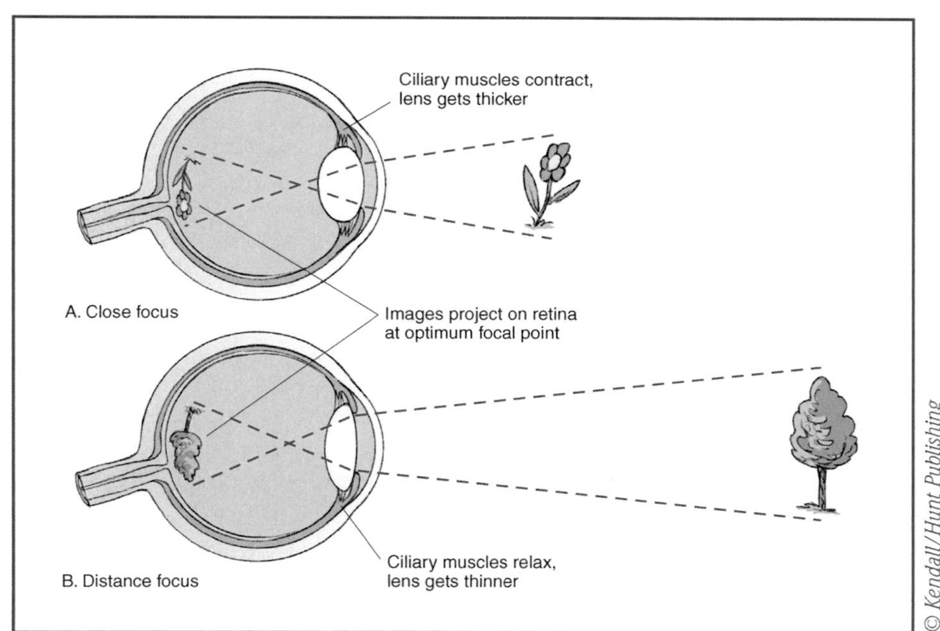

FIGURE 22.6

NAME ____

EXERCISE 22

QUESTIONS

1. What process is responsible for the similarity of brains across the animal kingdom?

2. Define or list the function of the following structures. Use your textbook if needed

   Choroid

   Optic nerve

   Round windows

   Eustacian tube

3. Look at Figure 22.1 and describe the differences between large and small mammalian brains.

4. How much would the pinna differ between two species in which one relied heavily on hearing and the other did not?

5. How would the eyes differ between two species where one was active during the day and the other at night?

EXERCISE 23

EMBRYOLOGY

INTRODUCTION

Embryology is the study of embryos. By studying the development of organisms you can look back into evolutionary history. The phrase "**ontogeny recapitulates phylogeny**" is appropriate here. Basically this means that the **ontogeny** (development) of an organism will show steps along the evolutionary path that this species took—in other words its **phylogeny**. If you examine the embryos of vertebrates you will notice that they are very similar in at least the early stages of development. It is also quite obvious that the more closely related the organisms are the more features they will have in common. In fact, if you compare the embryos of several vertebrates (including humans) you will be hard pressed to determine which one is the human. Only when more human-like characteristics become visible will you be able to tell for sure. By studying the embryos of various vertebrates we can gain insight into the development of our own species. That is the focus of today's lab exercise.

FROG EMBRYO

Figure 23.1 shows various stages of the development of the frog embryo, beginning with the unfertilized egg. Once fertilized, the cell begins to divide, but it does not increase in size. This division is called **cleavage**. At around the sixteen-cell stage the **morula** forms, which is simply a ball of cells. Next, the **blastula** forms. During this stage the cells arrange themselves into a ball of cells one layer thick with a fluid-filled cavity called the **blastocoel** in the center. Once the blastula forms, the cells begin to fold inward and form a two-layer **gastrula**. The cavity that forms produces a primitive gut with the opening called the **blastopore**. At this point we call the outer layer of cells the **ectoderm** and the inner layer the **endoderm**. Continued development of the gastrula will also produce a middle layer of cells called the **mesoderm**.

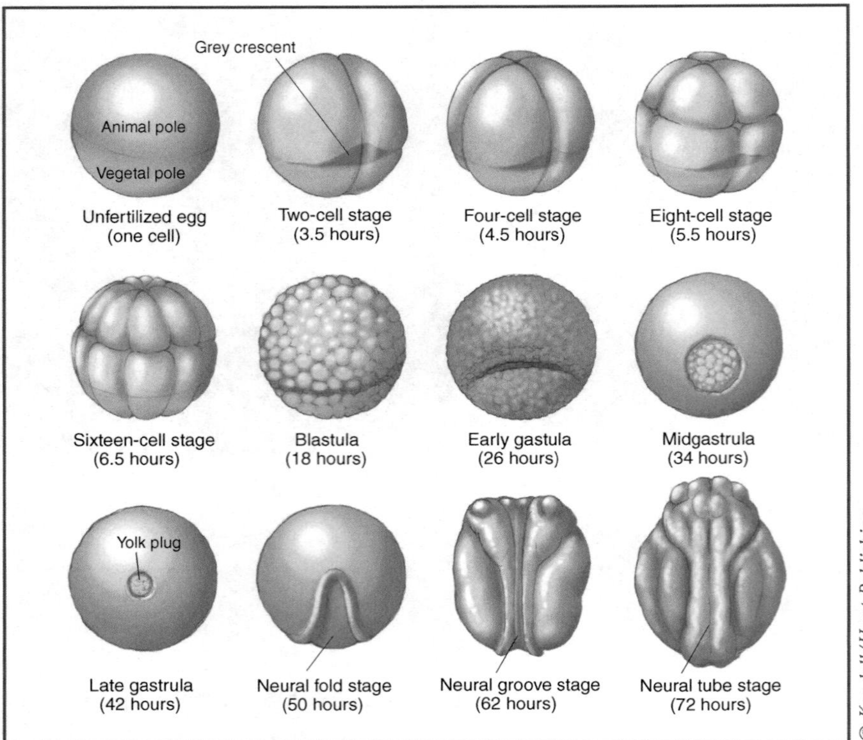

FIGURE 23.1

Obtain prepared slides of both frog and starfish developmental stages and locate the two, four, and eight-cells stages, the morula, the blastula, and the gastrula. Use the space below to draw what you see.

EMBRYONIC MEMBRANES

While frogs require an aquatic environment for reproduction, birds do not. Instead, birds have adapted a **hard amniotic shell**, which provides a suitable environment for the development of the embryo independent of the water. This is clearly an adaptation for a terrestrial existence. Placental mammals, like humans, have taken this one step further, and development of the embryo occurs within the body of the mother itself, aided by the **placenta.** Figure 23.2 shows the embryos of a bird and a mammal with the respective membranes indicated for each group. Compare the embryos of the bird and mammal and answer the questions below.

1. What is the difference in the yolk sac between birds and placental mammals?

2. What is the difference in the amnion between birds and placental mammals?

3. What is the difference in the allantois between birds and placental mammals?

4. Which embryo contains an umbilical cord?

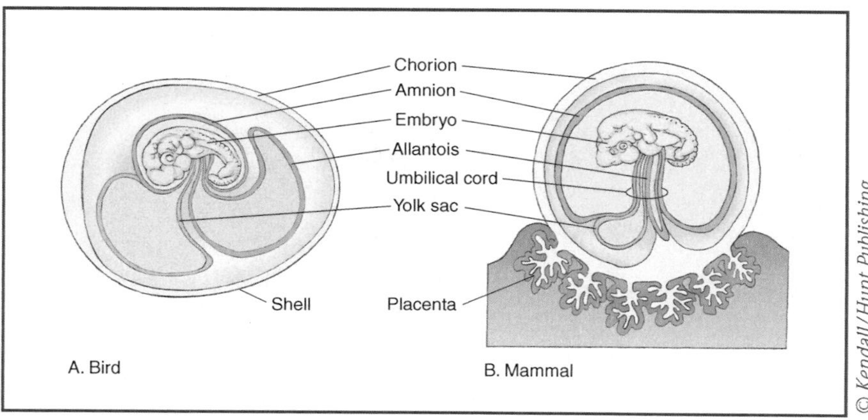

FIGURE 23.2

5. What is the connection between the placenta and the embryo?

6. Do you think the similarities between bird and placental mammal embryonic membranes suggest a common evolutionary history, and if so what is it? In other words, what did birds and mammals evolve from?

CHICKEN EMBRYOS

Figure 23.3 shows the development of the chicken embryo at 24, 48, 72, and 96 hours. Obtain prepared slides of the chicken embryo at these stages of development and compare those slides with Figure 23.3. Locate and identify all structures indicated on Figure 23.3.

1. At what point in the development of the chicken embryo does the brain begin to form?

2. At what point do the somites form?

3. When do the ears and the eyes form?

4. At what point do you see a heart begin to form?

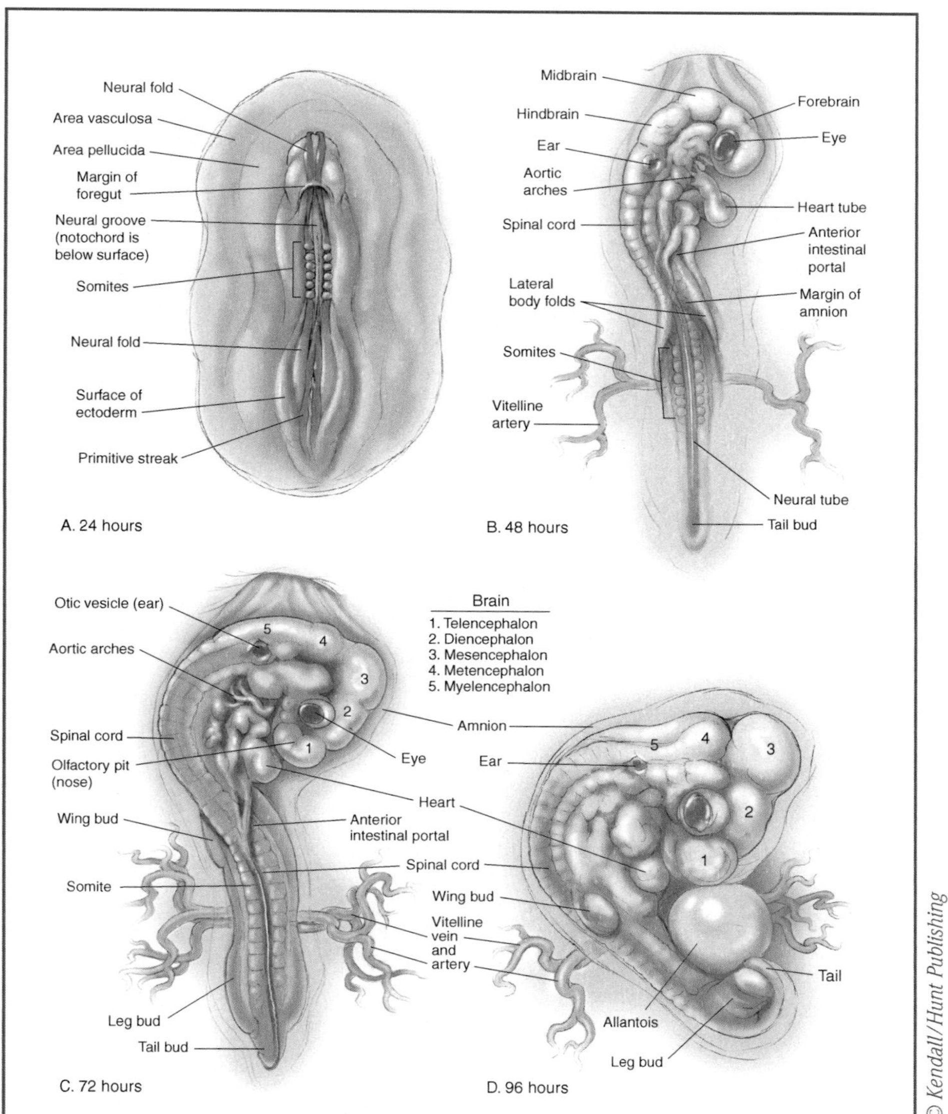

FIGURE 23.3

5. When do the wings begin to form?

6. When does the spinal cord develop?

Compare Figure 23.3 with models of human embryos provided by your instructor. What are the similarities and differences between human and chicken embryos?

NAME ______________________________________

EXERCISE 23

QUESTIONS

1. Define the following.

   Blastopore

   Blastocoel

   Endoderm

2. What is the difference in the yolk sac between birds and mammals?

3. Take a look at Figure 23.3 and answer the following.

   a. When do wings first form?

   b. When do the somites first form?

   c. When does the allantois first appear?

   d. When does the vitelline artery first appear?

4. What does "ontogeny recapitulates phylogeny" mean?

5. What is the connection between the placenta and embryo?

EXERCISE 24

DICHOTOMOUS KEYS

INTRODUCTION

Most people believe that biologists can identify all species on earth—after all since they study living things they should be able to identify all organisms, right? WRONG! While biologists can identify more species than the average person, there may be as many as fifty million species alive on earth today so there are more species that a biologist cannot identify than those he can identify. When the average person wants to identify species he will refer to a field guide that he can purchase at any local bookstore. These books, while appropriate for the lay naturalist, are not complete enough for serious biologists who must identify organisms to species. When biologists need to identify organisms they cannot readily recognize they will turn to **dichotomous keys**. Today you will learn to make and use a dichotomous key.

CHARACTERISTICS OF A DICHOTOMOUS KEY

It has been said that dichotomous keys are made by people who don't need them for people who cannot use them. This may sound foreboding, but dichotomous keys are simple in their design. A dichotomous key is an aid that will help to identify an organism by asking a series of yes or no questions about the organism in question. There will always be two choices, hence the name dichotomous key. The characteristics should be unambiguous and not vague. In other words, avoid characteristics like short and long. These characteristics are ambiguous because what is long to one person may be short to another. You can use characteristics like length provided you can measure length in some way. Instead of giving a characteristic like "tail long," you could give a characteristic like "tail longer than rest of the body." When constructing a key you will always give two choices. These choices will either take you to a set of two more choices or will identify the species. This may sound confusing, but it will become clear in the next section, which will give you a sample dichotomous key.

SAMPLE DICHOTOMOUS KEY

Examine Figure 24.1. This figure shows a variety of animals identified as follows, beginning in the upper left-hand corner and reading from left to right: pelican, chimney swift, pink flamingo, swan, ant, spider, grasshopper, snake, sawfish, shrimp (assume the shrimp has ten legs, five on each side), bivalve, and a clown fish. What would a dichotomous key designed to identify these animals look like? Look at the sample key below to find out.

1a. Animal has feathers . . . . Go to 2
1b. Animal does not have feathers . . . . Go to 5

2a. Lower bill makes a large scoop . . . . Pelican
2b. Lower bill not shaped like a large scoop . . . . Go to 3

3a. Tail shaped like a fork . . . . Chimney swift
3b. Tail not shaped like a fork . . . . Go to 4

4a. Feathers are pink . . . . Pink flamingo
4b. Feathers are white . . . . Swan

5a. Body elongated and with no lateral appendages . . . . Snake
5b. Body either not elongated or has lateral appendages . . . . Go to 6

6a. Body with six or more legs . . . . Go to 7
6b. Body with less than six legs . . . . Go to 10

FIGURE 24.1

7a. Hind legs much larger than first two pair . . . . .Grasshopper
7b. Legs are the same size . . . . .Go to 8

8a. Antennae curved backward toward head . . . . .Shrimp
8b. Antennae either not present or not curved backwards . . . . .Go to 9

9a. Three body segments . . . . .Ant
9b. Two body segments . . . . .Spider

10a. Has a well-developed head . . . . .Go to 11
10b. Does not have a well-developed head . . . . .Bivalve

11a. Saw-like structure is present . . . . .Sawfish
11b. No saw-like structure present . . . . .Clownfish

Notice that there are always two choices and that either you are directed to another set of choices or the species is identified. For instance, the first decision you have to make is whether or not the animal you need to identify has feathers or not. Depending on the answer to this question you will proceed to another set of choices until you have positively identified the organism in question.

PRACTICE DICHOTOMOUS KEYS

On a separate sheet of paper make a dichotomous key designed to identify the animals shown on Figure 24.2. The animals are as follows: a bat, rabbit, a rat, a chipmunk, and a squirrel. Remember to always give two choices and to either identify the species or send the reader to another set of choices.

Your instructor has provided you with a collection of small figures representing animals. Work as a group to make a dichotomous key for these animals. Do this on a separate sheet of paper.

FIGURE 24.2

Take a look at the front cover of this manual and construct a dichotomous key to identify the species shown.

EXERCISE 25

POPULATION MONITORING

INTRODUCTION

One of the challenges facing biologists when managing natural resources is determining the size of the population in question. This is especially important when considering whether or not to list a species as threatened or endangered. The problem with determining population sizes lies with the very nature of animals themselves. Typically animals are mobile, and many are secretive. These very qualities make determining population size very difficult. There are a number of methods for estimating population size such as listening to the numbers of birds singing, sighting animals, sighting tracks or other signs, or trapping. These methods only produce an estimate of the population at any given time. There is very little chance a biologist will know the exact number of animals within an area unless they are within an enclosed area, such as an island or an enclosed park, or if the animals are large and relatively visible, such as elephants. While you can use sightings of larger animals to estimate population, how do you go about estimating the population size of smaller animals such as rodents, which you cannot easily see or whose signs you may not easily find? That is the question we will address in today's lab by using a method called **capture/recapture.**

CAPTURE/RECAPTURE METHODS

Biologists use capture/recapture techniques to estimate sizes of animal populations. In 1930 biologists formulated the first capture/recapture method, called the Lincoln-Peterson estimator. They still use it today. The basic premise behind this method is very simple: Trap animals during two periods to estimate population size. During the first trapping period mark and release animals at the site of capture. At another point in time trap animals again. Record the total number of animals captured during both trapping periods as well as those also captured during the first trapping period. The formula for using this estimate appears below:

$$N = \frac{(n_1 n_2)}{m_2}$$

N = The population size estimate
n_1 = The number of animals captured and marked in the first sampling period.
n_2 = The number of animals captured during the second trapping period
m_2 = The number of animals captured during the second trapping period that were already marked. These were marked during the first trapping period.

Let's take a simple example to see how this method estimates population size. In this example you capture, mark, and release forty animals during the first time period. During the second trapping session you capture thirty animals total with two animals already bearing marks from the first trapping session. Using the above formula we get the following:

$$N = \frac{(40 \times 30)}{2}$$

$$N = 600$$

Based on the animals that you captured during each of the two trapping periods you estimate that there are six hundred individuals within the population in this area.

Suppose that instead of capturing two individuals already marked you captured fifteen. How would that influence your population estimate? In that case your estimate will be eighty individuals instead of six hundred. As you can see, the number of previously marked individuals greatly influences your estimate. This estimator has some biases associated with it that may influence the results derived. Because of this, biologists have modified the Lincoln-Peterson estimator to help eliminate some of these problems. The revised Lincoln-Peterson estimator appears below:

$$N = \left\{\frac{(n_1 + 1)(n_2 + 1)}{(m_2 + 1)}\right\} - 1$$

This appears more complicated, but it attempts to give a more realistic estimation of the population size compared to the original formula. Using data from the previous example let's use this formula to estimate the size of the population.

$$N = \left\{\frac{(40 + 1)(30 + 1)}{(2 + 1)}\right\} - 1$$

$$N = 422.7$$

$$N = 423$$

There are a number of assumptions with the Lincoln-Peterson estimator, which are:

1. The population being sampled is closed; the population size does not fluctuate between the two sampling periods.
2. The likelihood of capture for individual animals does not change between the two sampling periods. This means that animals do not become harder or easier to trap between trapping periods.
3. Marks are correctly recorded and do not fall off between sampling periods.

ESTIMATING THE POPULATION OF BEANS

Your instructor has provided you with a container of beans to represent the population of an animal you are going to estimate. You will do this using two "trapping" periods. For the first trapping period use a scoop or a small cup to remove a sample of the beans. This sample represents captures during the first trapping session. Use a pencil to mark the beans as if you are tagging an animal and release the beans back into their container. Mix the beans. This represents animal movement between the two trapping sessions. For the second trapping session use a scoop or a cup to remove another sample of beans. Do this without looking at the container to prevent scooping up marked or unmarked beans intentionally. Record the number of beans that you "captured" as well as any already marked. Use both the Lincoln-Peterson estimator and the revised version of the Lincoln-Peterson estimator to estimate the population. Use the space below to record your data and to work out the problem.

Repeat the above experiment four more times. Use different marks to avoid confusion with those captured during previous experiments. If you "capture" a bean used in a previous experiment just ignore that mark as if you had not captured it before. Use the space below to record your data and to estimate the population size using both formulas.

Once completed, count the actual number of beans in the container and record that number here. ____________

Answer the following questions:

1. Did your estimates between the different trials give the same results?

2. How did the results between the two formulas differ?

3. Based on the number of beans actually in the container, were your estimates accurate or inaccurate?

4. If they were inaccurate why use these estimates at all?

5. How realistic do you think the assumptions of the Lincoln-Peterson estimator are? Explain your answer.

6. How could you make sure the first assumption is met?

7. How could you make sure the second assumption is met?

NAME ____

EXERCISE 25

QUESTIONS

1. What are the formulas used to assess population size using capture/recapture methods?

2. What do you think the disadvantages and advantages are of using this method?

3. What are some of the other ways you can assess animal populations?

4. How does the number of recaptures influence population estimation using this method?

5. Why would you want to know the size of an animal's population?

6. Why would the size of an animal's population be difficult to determine?

EXERCISE 26

NATURAL SELECTION

INTRODUCTION

In 1831, a young biologist named Charles Darwin set upon one of the greatest voyages of discovery that the world has ever known. He was taken aboard the *HMS Beagle* as ship's naturalist. The *Beagle* spent 5 years exploring regions of South America, Africa, Australia, and the various islands encountered on this voyage. At this point in human history we did not know how organisms changed over time; however, we did know that organisms changed. How all this occurred remained a mystery. During this time, Darwin took notes and specimens from the areas he explored. All of this information led him to his ideas regarding the evolution of species. He outlined his ideas in a book entitled *On the Origin of Species by Means of Natural Selection* which was first published in 1859.

In this publication, which is now a classic in biology, he gave us the mechanism of evolutionary changed that he called "natural selection." **Natural selection** can be defined as the differential reproductive success of organisms due to heritable traits. Basically what this means is that there are traits that an organism possesses that give it an advantage and allows it to survive longer and successfully reproduce. These are **adaptive traits**. An adaptive trait is any trait that allows you to survive long enough to successfully reproduce. Additionally, these traits must have a genetic basis. That is, organisms with those traits must be able to pass that trait down to their offspring.

The way that natural selection works is very simple. Organisms with adaptive traits will be selected for, or favored, whereas those without those adaptive will be selected against. Organisms with adaptive traits will have greater reproductive success and those traits will build up within the population and those without the adaptive trait will diminish in numbers. As a result, the population will evolve to reflect those changes. This may seem confusing at first, but we are going to demonstrate this concept in today's lab.

QUESTIONS

When completing the questions below think in terms of natural selection. Think about the features that will be selected for; that is, those features that will give the organism an advantage. Use the space below the questions for your answers.

What type of traits would an animal need to survive hot/dry environments? What about extremely cold environments?

What type of traits would a predator need that preyed upon an animal that was heavily armored?

What type of defenses would a plant need to protect itself from a variety of herbivores? What features would the herbivores need to counter the plants defenses?

FIGURE 26.1

Take a look at Ffigure 26.1 and label the birds beginning in the upper left hand corner and moving clockwise A–G. Birds A and G are directly above and beneath each other. Answer the following questions.

- Which of these birds (there may be more than one) have adaptive features that would allow it to pry small insects from tight spaces? Why did you choose this/these bird(s)?

- Which of these birds (there may be more than one) do you think have adaptive features that would allow it to feed upon small seeds? Why did you choose this/these bird(s)?

- Which of these birds (there may be more than one) do you thing have adaptive features that allow if to feed upon fish? Why did you choose this/these bird(s)?

Design 2 species that are in a predator/prey relationship; that is, one species preys upon the other. List each species initial characteristics. Then list features that the prey would evolve to counter the predator. Then list the features the predator would evolve to get around the prey's adaptations. Carry the countermeasures out for 2 series.

	PREY SPECIES	PREDATOR SPECIES
Initial features		
Countermeasures		
Countermeasures		

Look at Figure 26.2. This feature shows 3 frames of a cheetah female while running. Cheetahs are the fastest land animal on Earth as they can easily sprint up to 60–70 mph. This is a real advantage for cheetahs as they regularly prey upon Thompson's gazelles which are also fast. What type of features do you think are favored in cheetahs that allow this type of speed?

FIGURE 26.2

DIRECTIONAL SELECTION

Directional selection occurs in a population when the frequency of traits moves in one direction. For instance, if a population changed from being 90% small and 10% large to 90% large and 10% small. The process of natural selection is behind the change. This is clearly demonstrated in the peppered moth. The peppered moth is found in England and North America and occurs in 2 color variations: light (Figure 26.3) and dark (Figure 26.4). Many years ago the light colored variant was the most common because they blended in with the color of the lichens on the trees (Figure 26.3). The dark colored variant was conspicuous and as a result they faced greater predation by birds. Then the tables were turned. The Industrial Revolution began and soot covered the lichens of the trees which darkened the trunks. The dark colored variant then had the adaptive trait and was selected for. The light colored moth was selected against because at this point it was conspicuous. In this situation the frequencies of the two traits changed from primarily light to primarily dark. We can demonstrate this with a simple game.

FIGURE 26.3

FIGURE 26.4

NATURAL SELECTION GAME

This game will demonstrate directional selection. You will start with 2 variations in a species. These variations can be anything you wish but for simplicity we will call them Variant 1 and Variant 2. Each variant consists of 20 individuals. You will need to construct a table similar to the table below.

	VARIANT 1—YEAR 1	VARIANT 2—YEAR 1
Number alive	20	20
Number of deaths		
Number remaining (Alive-Deaths)		
Number of breeding pairs (Number remaining/2)		
Number of births (Number of pairs/2)		
Number alive for next year (Number remaining+Births)		

We will make some assumptions. First let's assume that neither variant has an advantage. That is, neither variant is selected for or against when compared to the other variant. This will give us something to compare to future simulations where one variant is selected for. Also, let's assume that the breeding success of pairs is 50%. This means that if there are 10 breeding pairs there will be 5 births that will make it to adulthood for the next year. In reality the breeding success of pairs for many species is far less than 50%.

In this simulation we will have both variants equally likely to fall victim to a predator. You will need a standard 6-sided die for this game. Roll the die 10 times. If the number is between 1-3 that represents 1 individual from variant 1 killed. If the number is between 4-6 that represents 1 individual from variant 2 killed. Record this data in the "Number of Deaths" row. Let's do an example. After 10 rolls of the die there were 5 deaths for both variants. This is recorded and the other information is calculated.

	VARIANT 1—YEAR 1	VARIANT 2—YEAR 1
Number alive	20	20
Number of deaths	5	5
Number remaining (Alive-Deaths)	15	15
Number of breeding pairs (Number remaining/2)	14.5 (round down) = 14	14.5 (round down) = 14
Number of births (Number of pairs/2)	7	7
Number alive for next year (Number remaining+Births)	22	22

The number alive for year 2 is moved forward to a new table.

	VARIANT 1—YEAR 2	VARIANT 2—YEAR 2
Number alive	22	22
Number of deaths		
Number remaining (Alive-Deaths)		
Number of breeding pairs (Number remaining/2)		
Number of births (Number of pairs/2)		
Number alive for next year (Number remaining+Births)		

On a separate sheet of paper, continue rolling the die and recording results for a total of 5 years. If the number of births is not a whole number (i.e. 1.5) then round up to the next whole number (i.e. 2). Realize that if there is a situation where there is only 1 individual left there will be no breeding pairs. After 5 years, graph the results of the population sizes of the 2 variants and answer the following questions:

- Did either of the variants take over the population? Why?

- Did one variant seem to consistently have a higher rate of predation that the other variant? Why?

Let's now change the parameters of the game. Let's assume that variant 1 has an advantage over variant 2. In this case variant 1 has a camouflage pattern that is more like the environment in which it lives. In this case the die numbers are: a roll of 1 represents a death of an individual of variant 1 and a roll of 2-6 results in the death of a member of variant 2. Begin with 20 individuals as in the last simulation and continue until only 1 variant remains. Graph your results.

- In which direction did the frequency of traits shift? That is, which variant built up within the population?

Let's run this simulation again. Variant 1 has the advantage over variant 2 just like in the previous situation. However; after 4 years something changes where variant 2 has the advantage so the situation is reversed. After this change a roll of 1–5 represents a death of a member of variant 1 and a roll of 6 represents a death of a member of variant 2. Continue this until only 1 remains. If variant 2 dies off before the switch then backup and change the parameters at the point where there are still breeding pairs left in variant 2. Graph your results.

- What happened to the populations of variants 1 and 2 before and after the change?

Try this: Run any of the previous simulations that you would like; however, change some of the assumptions. For instance, you may want to change the breeding success of pairs or change the initial population size.

- How did your changes affect the outcome of the simulation compared to the previous simulations?

NAME ______________________________

EXERCISE 26

QUESTIONS

1. What is an adaptive trait?

2. In your own words define "natural selection."

3. What happens to individuals without adaptive traits?

4. When you compare humans to other animals we seem to be at a disadvantage. We are not very fast. We are not very strong. We cannot climb very well compared to other primates. Our sense of sight, smell, and hearing is very poor compared to other animals. How is it then that we are so successful? Do we have any adaptive traits, and if so what are they?